Sets: An Introduction

Sets
An Introduction

Michael D. Potter

Fellow and Director of Studies in Mathematics, Fitzwilliam College, University of Cambridge

Oxford · New York · Tokyo
CLARENDON PRESS
1990

Oxford University Press, Walton Street, Oxford OX2 6DP
Oxford New York Toronto
Delhi Bombay Calcutta Madras Karachi
Petaling Jaya Singapore Hong Kong Tokyo
Nairobi Dar es Salaam Cape Town
Melbourne Auckland
and associated companies in
Berlin Ibadan

Oxford is a trade mark of Oxford University Press

Published in the United States
by Oxford University Press, New York

British Library Cataloguing in Publication Data
Potter, Michael D.
Sets: an introduction
1. Set theory
I. Title
511.322

ISBN 0–19–853388–8
ISBN 0–19–853399–3 pbk

Library of Congress Cataloging in Publication Data
Potter, Michael D.
Sets: an introduction / Michael D. Potter.
p. cm.
1. Set theory. I. Title.
QA248.P68 1990 511.3′2—dc20 90–40303

ISBN 0–19–853388–8—ISBN 0–19–853399–3 (pbk.)

Set by
Asco Trade Typesetting Ltd, Hong Kong
Printed in Great Britain by
Biddles Ltd, Guildford and King's Lynn

To my Father

Preface

This is an introduction to set theory for mathematicians which I have been narrowly dissuaded from calling *The joy of sets.* The principal novelty in the treatment lies in the axiom system, which is a simplification of one proposed by Dana Scott and has not, I think, appeared in a textbook before, although its elegance and plausibility are strong reasons for preferring it to the usual ones. I have allowed the construction of the cumulative hierarchy to proceed beyond the universal class **D** of all sets, both as an aid to category theorists and as a curb on superstition. However, I do not assume the axiom of replacement. Nor have I chosen to outlaw individuals (sometimes called atoms): nothing hinges on their exclusion and it seems quite wrong to proscribe them without good reason.

In Chapter 4 I define the rational and real numbers only with respect to their orderings: to have defined their field structure properly would have taken the book further into the realm of conventional algebra than I wanted. I do not use the industry-standard von Neumann definition of ordinals: this is of course forced on me by the absence of the replacement axiom, but the resulting treatment is in any case rather neater than the usual one. Uses of the axiom of choice are signposted more clearly (and avoided more often) than in many elementary books. Finally, the treatment of the prime ideal property in Chapter 9 is fuller than in most textbooks and contains a couple of proofs not easily accessible elsewhere.

I am grateful to Richard Stamper, Peter Sullivan, John Derrick, Eric James, and my father for reading and commenting on parts of earlier versions of this book, and to Harry Simmons for giving me several years ago some lecture notes which influenced Chapter 8. I am also grateful to all the members of Balliol College and Lady Margaret Hall, Oxford who made the period when I wrote the book so stimulating.

Cambridge
January 1990

M.D.P.

Contents

Notation

Italic letters denote variables; roman letters denote constants. There is a convention among logicians that lower-case constants (called *proper names*) denote specific objects and upper-case ones (called *predicate letters*) denote specific properties which objects can have. In practice, though, it aids the legibility, allusiveness, and variety of a mathematical text to use as wide a range of typefaces and alphabets as possible*. We shall use the following notations from logic:

$\Rightarrow$ only if;
$\Leftrightarrow$ if and only if (also written 'iff');
$\forall$ for all;
$\exists$ for some;
$\exists\!\!\!\!\exists$ for some definite;
ι a definite;
$=$ equals;
! unique.

Linguistic expressions of the sort which can be used to make statements about objects are called *formulae*; expressions of the sort which can be used to denote particular objects are called *terms*.

Example Suppose that 'Nx' means 'x is a number' and 'Sxy' means 'y is the successor of x'. Then '$(\forall x)(\mathrm{N}x \Rightarrow (\exists! y)\mathrm{S}xy)$' is a formula meaning 'Every number has a unique successor'; and '$\iota! y\mathrm{S}xy$' is a term denoting the unique successor of x, i.e. $x + 1$.

This concludes the formal logic you need to know in order to understand the exposition of a first-order mathematical theory such as the one in this book. So if you neither know nor care what a first-order mathematical theory is, you may wish to omit the Introduction, which contains a few remarks about the features such theories have in common.

*The theory we shall be expounding in this book has only one (binary) predicate letter '$\in$' and one proper name '**D**': it thus immediately violates the logicians' convention.

Introduction: logic and language

In principle the exposition of a first-order mathematical theory need consist of no more than a list of assumptions, called *axioms*, followed by a list of assertions, called *theorems**, each of which is written in the language of the theory and follows from the axioms and the previous theorems by means of the rules of classical logic. Checking such a list of theorems for deductive correctness would be a very boring but entirely mechanical process and it would be easy to program a computer to do it. However, such a list would be quite useless as a means of communicating mathematics from author to reader and we shall not provide one here. What follows should be regarded instead as an informal specification of such a list: theorems will be stated in a mixture of the formal first-order language and informal English; we shall from the outset use *definitions* to introduce abbreviations, without which the statements of even the simplest theorems would be far too long to take in; and the 'proofs' will in fact be no more than attempts to indicate how a fully formalized proof could be constructed.

On formalism This characterization of our work as an informal specification of an entirely formalized text has two notable advantages. It rules out of court many potentially troublesome expressions which might otherwise be considered legitimate: for example, the expression 'the least natural number expressible as the sum of two cubes in two different ways' will be formalizable in our language, but 'the least natural number not definable in 25 or fewer syllables of English' will not. And it provides a clear criterion by which our proofs may be judged: a demonstration that an informal proof can be formalized in this way settles once and for all the question of its correctness.

Nevertheless, our dedication to formalism in this book is plainly of a rather watered-down, theoretical kind, since the process of encoding the (informal) arguments presented here in the (formal) first-order language of set theory is not an entirely mechanical one. Perhaps it is an indication of confidence in the security of our arguments that we can take such a relaxed view: compare the much more explicitly formal presentation in *Principia mathematica* (Whitehead and Russell 1910–13), for instance.

* We shall call some of the theorems in this book *lemmas*, *propositions*, or *corollaries*: the distinction is of no formal significance.

But the advantages of a more formal approach should not be dismissed too lightly. If we could develop a specification language for mathematics which was comprehensible to both humans and computers, we could then genuinely have our proofs checked mechanically for us: it seems likely that this would quite often detect human errors in the reasoning.

It is particularly valuable when we are presenting a part of mathematics whose interpretation is a matter of some controversy to separate strictly the pure (attempting to convince the reader that certain sentences are provable in the theory) from the applied (explaining the intended meaning of the sentences of the theory in a way which convinces the reader of the truth of the axioms and the resulting interest of their formal consequences). It has recently been common for mathematicians to concentrate on the pure part and leave the task of explaining the applied part to philosophers. This seems unwise.

First- and second-order theories We can conceive of there being a realm of things (called objects) capable of being denoted by names. In the same way, we might suppose that there is a realm of things which are capable of being denoted by formulae. If we did this, it would presumably be reasonable to introduce a new type of variable, written X, X', X'', etc., to denote these things. But it is by no means uncontroversial that such a realm exists, and we shall not suppose in this book that it does. If Quine is right that 'to be is to be the value of a variable' (1953, p. 15), it follows that in our formal language we ought not to permit quantification over anything other than objects; that is the content of the phrase 'first-order' which we have already used in describing our theory.

The self-denying ordinance which limits our formal language to first-order quantification in this way is not without its penalties. In particular, the theory we shall be studying in this book is sufficiently rich in structure that it cannot be conveniently based on a finite list of first-order axioms. We shall therefore have to allow rules called *AXIOMS* (also known as 'axiom schemes') which specify an infinite list of axioms. If we demand that each of these rules be clearly transformable into a criterion for deciding mechanically whether a given string of symbols is an axiom in the list or not, then the claim that the formal text we are describing is mechanically checkable will not be affected by the fact that the list of axioms is not finite.

If you do not have any doubts about the reality of the things which formulae denote, the device of using AXIOMS will seem pointless and artificial; this will in particular be the case if your belief in the validity of an AXIOM stems entirely from a belief in the second-order axiom of which it represents a list of instantiations. There are nevertheless two contrasting technical points to be made: on the one hand, results in the meta-theory of first-order logic—the Löwenheim/Skolem theorem, nonstandard models, etc.—tell us how limited its expressive power is, and there is no doubt that this is at first sight surprising; but the exposition which follows shows, conversely, how much can be expressed in a first-order theory—broadly speaking, the whole of mathematics—and this is also genuine information. If there is any sleight of hand involved in trying to synthesize second-order structure by means of a first-order scheme (in our case, the AXIOM OF SUBCOLLECTIONS in §1.3), then it is at least more honest to do it in full view than to slip in so much commitment by the logical back door.

When we need to comment on the features that certain formulae have in common, or to introduce uniform abbreviations for formulae with a certain pattern, we shall find it helpful to use upper-case Greek letters such as Φ, Ψ, etc. to represent arbitrary formulae; if the formula Φ depends on the variables $x_1, \ldots, x_n$ then we can write it $\Phi(x_1, \ldots, x_n)$ to highlight that fact.

Remark 1 These Greek letters are to be thought of as part of the specification language (metalanguage) which we use to describe the formal text, not as part of the formal language itself. Their use in this schematic manner is ontologically innocent.

Example If $\Phi(x)$ is a formula then the formula

$$(\forall y)(\Phi(y) \Leftrightarrow x = y)$$

is written $\Phi(!x)$ and read 'Φ holds uniquely of x'. The formula $(\exists x)\Phi(!x)$ is also written $(\exists !x)\Phi(x)$ and read 'There exists a unique x such that $\Phi(x)$'.

Remark 2 The formula $(\forall y)(\Phi(y) \Leftrightarrow x = y)$ depends on the variable x; logicians say that x occurs 'free' in it. However, it does not depend on y: y is simply a dummy—logicians say 'bound'—variable and could be replaced by any other variable (except x) without affecting the intended meaning of the expression. (Compare the use of x in $\int_0^1 f(x)\,dx$.)

Remark 3 Strictly speaking, our definition of $\Phi(!x)$ is unsatisfactory as it stands. If $\Phi(x)$ were the formula '$x = y$', for example, then we would find that $\Phi(!x)$ is an abbreviation for $(\forall y)(y = y \Leftrightarrow x = y)$, which is true iff x is the only object in existence, whereas we intended it to mean 'x is the unique object equal to y', which is true iff $x = y$. What has gone wrong is that the variable y occurs in this choice of the formula $\Phi(x)$ and has therefore become accidentally bound. This sort of collision of variables is an irritating feature of quantified logic which we shall ignore from now on. To be explicit, we shall assume whenever we use a variable that it is chosen so that it does not collide with any of the other variables we are using: this is always possible since the number of variables occurring in any given formula is finite whereas the number of variables we can create is unlimited. (We can add primes to x indefinitely to obtain x', x'', etc.)

Remark 4 Bradman was the greatest batsman of his time; 'Bradman' is a name with seven letters. Confusion between names and the objects which they denote (or between formulae and whatever it is that *they* denote) can be avoided by such careful use of quotation marks. Another strategy, which we shall adopt in this book, is to rely on the reader's common sense to achieve the same effect.

If $\Phi(x)$ is a formula, then the expression $\iota x\Phi(x)$ is read 'a definite x such that $\Phi(x)$'. Expressions of this form are called *descriptions*. (The words 'canonical' and 'explicit' are sometimes used here instead of 'definite'.) If $\Phi(x, x_1, \ldots, x_n)$ is a formula depending on the variables $x, x_1, \ldots, x_n$, then $\iota x\Phi(x, x_1, \ldots, x_n)$ is a term depending on $x_1, \ldots, x_n$. Proper names are examples of terms not depending on any variables. We shall use lower-case Greek letters such as σ, τ, etc. to represent arbitrary terms; if the term σ depends on the variables $x_1, \ldots, x_n$, then we can write it $\sigma(x_1, \ldots, x_n)$ to highlight that fact.

Remark 5 These schematic lower-case Greek letters are, like the upper-case letters we introduced earlier, part of the specification language, not of the object language: they stand in a schematic sentence in the places where particular terms stand in an actual sentence.

If a term 'σ' denotes something, then we shall say that σ *exists*. It is a convention of language that proper names always denote something. The same is not true of descriptions: consider for example the description $\iota x(x \neq x)$. Instead of '$\iota x\Phi(x)$ exists' we shall sometimes write $(\exists\!\!\!\exists x)\Phi(x)$, which is read 'there exists a definite x such that $\Phi(x)$'. We adopt the convention that if σ and τ are terms, the equation $\sigma = \tau$ is to be read as meaning 'if either σ or τ exists then they both do and they are equal'.

We certainly intend that definite existence should imply existence, i.e. that if $\Phi(x)$ is a formula then

$$(\exists\!\!\!\exists x)\Phi(x) \Rightarrow (\exists x)\Phi(x).$$

It would evidently simplify matters considerably if we were to remove the distinction between existence and definite existence completely and assume that

$$(\exists x)\Phi(x) \Rightarrow (\exists\!\!\!\exists x)\Phi(x).$$

This assumption is called the *AXIOM OF STRONG CHOICE*. It is a rather extreme expression of a non-constructive viewpoint in logic since it conflates the distinction between knowledge that there are objects with some property, which may derive from a highly indirect (e.g. impredicative) existence proof, and the exhibiting of an explicit term which satisfies the property. We shall therefore not assume it in this book. Instead we restrict ourselves to two special cases.

The first (and less controversial) is when there is a *unique* x such that $\Phi(x)$. The term $\iota x\Phi(!x)$ will also be written $\iota!x\Phi(x)$ and read '*the* x such that $\Phi(x)$'. The formula $\exists x\Phi(!x)$ will also be written $\exists!x\Phi(x)$; it is evidently equivalent to '$\iota!x\Phi(x)$ exists'. What we have to assume as an axiom is that if $\Phi(x)$ is a formula, then

$$(\exists!x)\Phi(x) \Rightarrow (\exists!x)\Phi(x).$$

The second (more controversial) usage of the term $\iota x\Phi(x)$ occurs whenever we can exhibit a term τ such that $\Phi(\tau)$ holds and such that τ does not depend on any variables apart from those (other than x) on which $\Phi(x)$ depends. This has the effect of allowing us, if we have exhibited a term τ such that $\Phi(\tau)$, to choose such a term once and for all and then forget (formally at any rate) which one it was that we chose. (We shall not use this more controversial sort of term until §4.1, where the note 'The use of abstraction terms' discusses it further.) To be explicit, if τ is a term and $\Phi(x)$ is a formula which depends on no variables other than x except possibly those on which τ depends, then we assume that

$$\Phi(\tau) \Rightarrow (\exists x)\Phi(x)$$

Remark 6 There is one property which is worth singling out: if $\Phi(x)$ and $\Psi(x)$ are formulae, then

$$(\forall x)(\Phi(x) \Leftrightarrow \Psi(x)) \Rightarrow \iota!x\Phi(x) = \iota!x\Psi(x).$$

A desire for tidiness might lead us also to assume (*IOTA CONSISTENCY AXIOM*) that if $\Phi(x)$ and $\Psi(x)$ are formulae which depend on the same variables, then

$$(\forall x)(\Phi(x) \Leftrightarrow \Psi(x)) \Rightarrow \iota x\Phi(x) = \iota x\Psi(x).$$

On definitions Because definitions are formally just ways of introducing abbreviations, the question of their correctness is simply one of whether they enable us mechanically and unambiguously to eliminate the expression being defined from every formula in which it occurs; the correctness in this sense of the definitions in this book will always (I hope) be trivially apparent. (The question of their psychological potency is of course quite a different matter.)

The things for which definitions introduce abbreviations will be either formulae or terms. If they are terms then apart from their formal correctness (i.e. unambiguous eliminability) there is the question of their existence. It is not wrong to use terms which do not denote anything, but it may be misleading since the rules of logic are not the same for them as they are for ones which do. (For example, the move from $\Phi(\sigma)$ to $(\exists x)\Phi(x)$ is wrong if σ does not exist.) So the introduction of a new symbol to abbreviate a term will often be followed by a *justification* to show that this term denotes something; if it is not followed by one, the reason is probably that the justification in question is completely trivial.

One more piece of terminology. A term (x, y) is called a *pairing term* in x and y if it depends only on x and y and

$$(\forall x)(\forall y)(\forall z)(\forall t)((x, y) = (z, t) \Leftrightarrow (x = z \text{ and } y = t)).$$

If (x, y) is a pairing term, it is traditional to write (x, y, z) for $((x, y), z)$, (x, y, z, t) for $((x, y, z), t)$, etc.

1 Collections

In every part of mathematics we study objects. Whether these objects be numbers, points in space, or elements of a locally convex topological vector space, they can be formed into collections. In this chapter (and indeed throughout the book) we shall study the properties which collections have simply by virtue of being *collections*, no matter what they are collections *of*. In other words, we shall be concerned only with properties arising from the relationship which holds between an object and any collection to which it belongs: this relationship is called membership.

1.1 Membership

The fundamental relationship between objects, which we take to be primitive, is that of *membership*: if x and y are objects, we write $x \in y$ (or occasionally $y \ni x$) to indicate that x is a member of y; we may then say that x is an *element* of y or that it *belongs* to y. We write $x \notin y$ as an abbreviation for 'not $(x \in y)$'.

Definition

If $\Phi(x)$ is a formula then the term $\iota! y (\forall x)(x \in y \Leftrightarrow \Phi(x))$ is abbreviated $\{x : \Phi(x)\}$ and read 'the collection of all x such that $\Phi(x)$'.

In words: $\{x : \Phi(x)\}$ is the unique object whose elements are precisely the objects satisfying Φ.

Lemma 1.1.1

(a) If $\Phi(x)$ is a formula such that $a = \{x : \Phi(x)\}$ exists, then $(\forall x)(x \in a \Leftrightarrow \Phi(x))$.

(b) If $\Phi(x)$ and $\Psi(x)$ are formulae such that $(\forall x)(\Phi(x) \Leftrightarrow \Psi(x))$, then $\{x : \Phi(x)\} = \{x : \Psi(x)\}$.

Proof (a) This follows at once from the definition.

(b) If $(\forall x)(\Phi(x) \Leftrightarrow \Psi(x))$, then

$$(\forall x)(x \in y \Leftrightarrow \Phi(x)) \Leftrightarrow (\forall x)(x \in y \Leftrightarrow \Psi(x)),$$

and so $\{x : \Phi(x)\} = \{x : \Psi(x)\}$ (cf. Introduction, remark 6). □

But when does $\{x : \Phi(x)\}$ exist? It would clearly be counter-intuitive to suppose that it always does, and indeed such a supposition rapidly leads to a contradiction:

Proposition 1.1.2 (*Zermelo/Russell paradox*)
$\{x : x \notin x\}$ does not exist.

Proof Suppose that $a = \{x : x \notin x\}$ exists. Then

$$(\forall x)(x \in a \Leftrightarrow x \notin x) \quad \text{[lemma 1.1.1(a)]}.$$

Therefore in particular $a \in a \Leftrightarrow a \notin a$. Contradiction. □

Collection-as-one; collection-as-many A paradox is a fact which is contrary to expectation (from the Greek παρά + δόξα, 'beyond expectation'). Whether one finds proposition 1.1.2 paradoxical depends on the meaning one attaches to the notion of a collection. According to the point of view we shall adopt in this book a collection can be thought of as a sack (the braces { }) containing its elements. This conception inevitably lends a hierarchical structure to our system: given some objects we can form new objects which collect together some of the old objects; then we can collect these new objects into further new objects; and so on. Therefore proposition 1.1.2 is unsurprising: on this view a collection plainly cannot belong to itself, so $\{x : x \notin x\}$ would, if it existed, be the collection of all collections, which is an absurdity. This conception ('collection-as-one') has several other important consequences: a singleton $\{x\}$ (see §1.3 below) is a different object from x itself; membership and inclusion are quite distinct notions; and the empty collection is perfectly comprehensible (think of an empty sack). However, the view that was common among early writers on set theory was rather different. According to this second view ('collection-as-many') a collection consists directly of its members (no braces). Hence in this conception x is precisely the same thing as the collection whose only element is x; membership is a special case of inclusion; and the notion of the empty collection loses its meaning. As Frege memorably expressed it in his review of Schröder, a collection-as-many 'consists of objects; it is an aggregate, a collective unity, of them; if so it must vanish when these objects vanish. If we burn down all the trees of a wood, we thereby burn down the wood. Thus there can be no empty class.' (Frege 1895, pp. 436–7.)

The collection-as-one conception seems to have originated with Cantor and became dominant among mathematicians in the early years of this century. More recently the collection-as-many conception has been pursued only by Lesniewski (see Fraenkel *et al.* 1973, pp. 200ff.), and his work has been almost totally ignored. Nevertheless at the end of the nineteenth century it was this notion which was common. Dedekind, for example, plainly had it in mind in *Was sind und was sollen die Zahlen?* (1888) when he avoided the empty collection and used the same symbol for membership and inclusion: he drafted an emendation adopting the collection-as-one position only much later (see Sinaceur 1971).

The credit for distinguishing between membership and inclusion is generally given to Peano, who introduced different notations for the two concepts and warned against confusing them in his publication of 1889. None the less, a couple of pages later he asserted that (in modern notation)

$$k \subseteq s \Rightarrow (k \in s \Leftrightarrow (k \neq \varnothing \text{ and } (\forall x, y \in k)(x = y)))$$

(in words: a subcollection k of s is also an element of s iff it has exactly one element), which makes precisely the blunder he had warned against. In 1890 he introduced a notation to distinguish between a collection b and the singleton $\{b\}$ (which he denoted ιb), but his motivation for this was somewhat quaint (1890, p. 192): 'Let us decompose the sign = into its two parts *is* and *equal to*; the word *is* is already denoted by $\in$; let us also denote the expression *equal to* by a sign, and let ι (the first letter of ἴσος) be that sign; thus instead of $a = b$ one can write $a \in \iota b$.'

Frege and the paradoxes Frege based his treatment of classes on the notion of 'concept': a concept is the thing (whatever it is) that materially equivalent predicates have in common. Predicates denote concepts in the same way that proper names denote objects: different proper names can denote the same object; and materially equivalent predicates can denote the same concept. Thus 'Lewis Carroll' and 'Charles Dodgson' denote the same man; and 'x is a prime number less than 10' and 'x is 2, 3, 5, or 7' denote the same concept. But concepts are not objects, or at any rate they are not objects of the same type as the objects falling under them: if we start with objects which we regard as entities of level zero, then concepts are first-level entities and so anything predicable of such a concept denotes a *second-level* concept; and so on. If we are to regard totalities in this intensionally derived way, we must therefore develop a hierarchical logic of some complexity (see Russell 1908). Frege avoided this by supposing that a concept of any level determines a ground-level totality (object) which he called its 'extension', thus collapsing the type distinctions we have referred to.

But this amounts to assuming that $\{x : \Phi(x)\}$ exists for every formula $\Phi(x)$, which is contradictory. The story of how Russell informed Frege of the irremediable flaw in his theory when the second volume of his *Grundgesetze der Arithmetik* was already at the printer is both dramatic and well known.

The historical origin of this paradox lies in a theorem Cantor proved in (1892) (theorem 5.1.4 here), a trivial consequence of which is that there is no collection a such that $(\forall b)(b \subseteq a \Rightarrow b \in a)$, whereas the collection of all collections would, if it existed, clearly have this property. Russell obtained this paradox and by analysing its proof got the one which generally carries his name some time in 1901 (see, Russell 1944, p. 13); Zermelo had obtained it independently in 1900 or 1901 (see Rang and Thomas 1981). It was not the first paradox to be discovered, however: Cantor (letter to Hilbert, 1897*b*) wrote that 'the collection of all alephs ... cannot be interpreted as a definite, well-defined finished set'. Nor was the sad appendix which Frege added to the second volume of the *Grundgesetze* the first reference in print to a paradox of this kind: Hilbert's (1900) lecture on the problems of mathematics referred to 'the system of *all* cardinal numbers or even of *all* Cantor's alephs, for which, as may be shown, a consistent system of axioms cannot be set up ...'.

What is particularly striking about the Zermelo/Russell paradox is its simplicity: we have proved it without having yet stated any axioms for our theory, whereas all the other set-theoretic paradoxes known at that time involved cardinal or ordinal numbers in some way. Curiously, Hilbert described these other paradoxes as 'even more convincing' (letter to Frege, 1903); it is not clear why. (Hilbert's first response to Cantor's paradox was apparently disbelief: compare Cantor's tactful reply (letter to Hilbert, 1897*c*).)

One step back from disaster To summarize: the so-called paradoxes of set theory arose not from an inconsistency in our intuitive notion of set but from a conflation of two or more incompatible notions (set-as-one, set-as-many, etc.). This view, although common now, is not often to be found expressed in the literature before the 1950s. What was much more common then was to view the paradoxes as exhibiting a genuine contradiction in our intuitive conception. 'The attitude is frankly pragmatic; one cures the visible symptoms [of the paradoxes] but neither diagnoses nor attacks the underlying disease.' (Weyl 1949, p. 231.) Quine put the point even more starkly (1941, p. 153): 'Common sense is bankrupt, for it wound up in contradiction. Deprived of his tradition, the logician has had to resort to myth-making.'

According to this view, the object of a good axiomatization is to retain as many as possible of the naive set-theoretic arguments which we remember with nostalgia from our days in Cantor's paradise, but to stop just short of permitting those arguments which lead to paradox. This certainly appears to have been the motive of Zermelo's axiomatization: 'There is at this point nothing left for us to do but to proceed in the opposite direction and, starting from set theory as it is historically given, to seek out the principles required for establishing the foundations of this mathematical discipline. In solving this problem we must, on the one hand, restrict these principles sufficiently to exclude all contradictions and, on the other, take them sufficiently wide to retain all that is valuable in this theory.' (Zermelo 1908*a*, p. 261.) In other words, absence of contradiction is to be regarded 'as an empirical fact rather than as a metaphysical principle' (Bourbaki 1949*b*, p. 3). There was a time when it was popular to regard this gung-ho pragmatism as an inevitable concomitant of platonism: 'The platonist can stomach anything short of contradiction; and when contradiction does appear, he is content to remove it with an *ad hoc* restriction.' (Quine 1953, p. 127.) But this characterization of platonism as a position without principles seems a trifle unfair.

The most pressing difficulty for the pragmatic view is that its security seems to depend on no more than its failure so far to lead to contradiction. Bourbaki (1954, p. 8): 'During the 40 years since we have formulated with sufficient precision the axioms of [set theory] and drawn their consequences in the most varied branches of mathematics, we have never come across a contradiction, and we are entitled to hope that one will never be produced.' Of course, it is impossible to deny that the time (now 80 years) which has elapsed without a contradiction being found is psychologically influential in engendering confidence in the system. But this can scarcely be regarded as amounting to very much. The claim that the system is formally consistent is in principle refutable simply by exhibiting a proof of a contradiction in it. But mathematicians routinely use only a tiny fragment of the generality permitted by the theory and it would presumably only be by pushing the theory to its limits that a contradiction could be obtained. So the lapse of time can only contribute significantly to confidence if attempts are being made to produce such a refutation: I know of no such attempts. (It may be significant that the position is different for certain other theories, notably Quine's NF and ML, which lack the positive intuitive underpinning which our theory has, although the emphasis on syntactic analysis in the studies which have been undertaken of Quine's systems is no doubt a consequence of the syntactic paradox-barring which motivated them.)

1.2 Collections

We have defined 'collection of ...', but not 'collection' (although we have already used the notion informally in the notes). The naive idea is that collections are precisely objects of the form $\{x : \Phi(x)\}$ for some formula $\Phi(x)$. This does not work under the formal restrictions we imposed in the Introduction (since the phrase 'for some formula Φ' is not first order). Instead:

Definition

We say that a is a *collection* if $a = \{x : x \in a\}$.

In the remainder of this chapter a, b, c, a', b', c', etc. will always denote collections.

In particular the quantifiers $(\forall a)$ and $(\exists a)$ should be read 'For every collection a' and 'For some collection a' respectively.

Lemma 1.2.1

Suppose that $\Phi(x)$ is a formula.
(a) If $\{x : \Phi(x)\}$ exists, then it is a collection.
(b) $(\exists a)(\forall x)(x \in a \Leftrightarrow \Phi(x))$ iff $\{x : \Phi(x)\}$ exists.

Proof (a) If $a = \{x : \Phi(x)\}$ exists, then

$$(\forall x)(x \in a \Leftrightarrow \Phi(x)) \quad \text{[lemma 1.1.1(a)]},$$

so that $\{x : x \in a\}$ exists and

$$a = \{x : \Phi(x)\} = \{x : x \in a\} \quad \text{[lemma 1.1.1(b)]}.$$

(b) If a is a collection such that $(\forall x)(x \in a \Leftrightarrow \Phi(x))$ then

$$a = \{x : x \in a\} = \{x : \Phi(x)\} \quad \text{[lemma 1.1.1(b)]}$$

and so $\{x : \Phi(x)\}$ exists. Conversely, if $a = \{x : \Phi(x)\}$ exists then it is a collection [(a)] and $(\forall x)(x \in a \Leftrightarrow \Phi(x))$ [lemma 1.1.1(a)]. □

Indefinable collections So collections are indeed precisely the objects of the form $\{x : \Phi(x)\}$. But in order to say this we have to allow that the formula Φ may involve the collection which it defines: the collection a is defined by the formula $x \in a$. Such a 'definition' is plainly circular and gives no new information about a. Could there be collections which are not, even in principle, genuinely definable? Explicitly, could there be a collection a which is not of the form $\{x : \Phi(x)\}$ for any formula Φ not depending on a? This is problematic. The view one takes seems to be rather directly dependent on what one thinks about the ontological and epistemological status of collections.

The passage from Peano quoted in §1.1 is one among many which suggest that he regarded talk about totalities—he called them 'classes'—as merely an alternative (extensional) way of expressing talk about (intensional) predicates. This view—the 'no-class theory'—was explicitly enunciated by Russell (1906): he proposed to rewrite assertions about classes so that all references to classes disappear. This frees the theory from ontological commitment and loads all the responsibility on the underlying logic. It therefore holds evident attractions for the logicist. This view is known as *nominalism*.

Of course, one can believe that talk of indefinable collections is incoherent without believing that the very notion of a collection is an eliminable fiction. This position might be called *cautious realism*. Cantor (1882, p. 114) explicitly required that for a set of elements to be well defined 'it must be seen as intrinsically determined on the basis of its definition and as a consequence of the logical principle of excluded middle whether any object ... belongs to the imagined manifold as an object or not ...'. By 1895–7 Cantor had dropped the requirement that a collection should have a definition but he never explicitly allowed the possibility that it might not have one. Dedekind went as far as to remark that 'in what manner this determination [whether an element belongs to the collection] is brought about, and whether we know a method of deciding upon it, is a matter of indifference for all that follows' (1888, No. 2, note), but this stops some way short of endorsing indefinable collections.

It was not until 1904 that Zermelo's explicit enunciation of the axiom of choice, which asserts the existence of certain collections without providing a means for defining them, brought the issue to prominence. Since then the dominant view among mathematicians has been that collections exist in extension quite independent of the language we have at our disposal for defining them. Thus, for example, Ramsey (1926, p. 355): 'The classes of male and female angels may be infinite and equal in number, so that it would be possible to pair off completely the male with the female, without there being any real relation such as marriage correlating them. The possibility of indefinable classes and relations in extension is an essential part of the extensional attitude of modern mathematics....'

To hold this view it is necessary (but perhaps not sufficient) to hold a *strong realist* (or platonist) conception of mathematics such as the one most famously expressed by Gödel (1944, p. 137): 'Classes ... may ... be conceived as real objects, namely ... as "pluralities of things" or as structures consisting of a plurality of things .... It seems to me that the assumption of such objects is quite as legitimate as the assumption of physical bodies and there is quite as much reason to believe in their existence. They are in the same sense necessary to obtain a satisfactory system of mathematics as physical bodies are necessary for a satisfactory theory of our sense perceptions.'

Proposition 1.2.2 (*Extensionality principle*)

$$(\forall x)(x \in a \Leftrightarrow x \in b) \Rightarrow a = b.$$

Proof Suppose that a and b are collections. Then $a = \{x : x \in a\}$ and $b = \{x : x \in b\}$. But if $(\forall x)(x \in a \Leftrightarrow x \in b)$, then

$$\{x : x \in a\} = \{x : x \in b\} \quad \text{[lemma 1.1.1(b)]},$$

and so $a = b$. □

In words: a collection is entirely determined by its elements.

Extensionality The extensionality principle is taken as an axiom in Zermelo (1908*a*)—where it is called the axiom of definiteness (*Axiom der*

Bestimmtheit)—and in most treatments since. The presentation we hav given here emphasizes its purely definitional character: a collection is th sort of thing which is determined by its elements. But our proof c extensionality depends on the properties of the description operator which we introduced in the Introduction, and the work of Robinso (1939) and Scott (1962) suggests that this dependence is essential.

Definition

An object which is not a collection is called an *individual.*

Remark It would be reasonable to add an axiom asserting that individuals hav no elements, but in practice we can—by dint of only a little sleight o hand—do without it.

1.3 Subcollections

The formula $(\forall x)(x \in a \Rightarrow x \in b)$ is abbreviated $a \subseteq b$ and read 'a is con tained in b' or 'b contains a' or 'a is a subcollection of b'; the formul '$a \subseteq b$ and $a \neq b$' is abbreviated $a \subset b$ and read 'a is strictly contained i b' or 'a is a proper subcollection of b'.

We cannot form collections arbitrarily. However, if we can form a col lection a then we can evidently also form the subcollection of a consistin of those members which share some particular property:

AXIOM OF SUBCOLLECTIONS

If $\Phi(x)$ is a formula, then the following is an axiom:

$$(\forall a)(\{x : x \in a \text{ and } \Phi(x)\} \text{ exists}).$$

We often write $\{x \in a : \Phi(x)\}$ for $\{x : x \in a \text{ and } \Phi(x)\}$.

The axiom of subcollections An informal version of this axiom appear in Zermelo (1908*a*), where it is called the 'axiom of separation' (*Axiom de Aussonderung*). But Zermelo did not define his object language precisel enough to exclude clearly the semantic paradoxes. Weyl attempted definition in 1910 but his work was ignored and the gap was refilled b Skolem in 1922.

We tried above to give the impression that the acceptability of th AXIOM OF SUBCOLLECTIONS is self-evident, and from a naiv perspective it surely is. However, this Panglossian view is misleading: th problems which it raises are among the most intractable in the philosoph of set theory. (Compare 'Predicative set theory' (§1.5), 'Need a number theorist believe Goodstein's theorem?' (§6.6), and 'Undecidable sentences and second-order axioms' (§7.5).)

It would be good if we could now prove the existence of a few collections. But the AXIOM OF SUBCOLLECTIONS is insufficient for this. We need to assume the existence of at least one collection. (In Chapter 3 we shall make a stronger assumption: hence the 'temporary' label.)

Temporary axiom

There exists a collection.

Now we can prove that there is exactly one collection with no elements:

Definition

$\varnothing = \iota! b(\forall x)(x \notin b)$ ('the *empty* collection').

Existence Suppose that a is a collection [temporary axiom]. Then $b = \{x \in a : x \neq x\}$ exists [AXIOM OF SUBCOLLECTIONS]. So

$$x \in b \Leftrightarrow (x \in a \text{ and } x \neq x) \Leftrightarrow \text{contradiction},$$

i.e. $(\forall x)(x \notin b)$.

Uniqueness Suppose b and b' are both collections with no elements. Then

$$x \in b \Leftrightarrow \text{contradiction} \Leftrightarrow x \in b'.$$

So $b = b'$ [proposition 1.2.2]. □

Definition

We introduce the following abbreviations:

(a) $\bigcap a = \{x : (\forall b \in a)(x \in b)\}$ (provided that a has at least one element c) ('intersection of a')

(b) $a \cap b = \{x : x \in a \text{ and } x \in b\}$

(c) $a - b = \{x : x \in a \text{ and } x \notin b\}$ ('*relative complement* of b in a')

Justification (a) If $c \in a$, then

$$a' = \{x \in c : (\forall b \in a)(x \in b)\}$$

exists [AXIOM OF SUBCOLLECTIONS]. But then

$$x \in a' \Leftrightarrow (x \in c \text{ and } (\forall b \in a)(x \in b)) \Leftrightarrow (\forall b \in a)(x \in b)$$

since $c \in a$. So $\{x : (\forall b \in a)(x \in b)\}$ exists [lemma 1.2.1(b)].

(b) and (c) Trivial [AXIOM OF SUBCOLLECTIONS]. □

Remark 1 Two collections a and b are said to be *disjoint* if $a \cap b = \varnothing$.

Remark 2 Note that $\bigcap \varnothing$ does not exist, since if it did it would be the collection of all collections and $\{x : x \notin x\}$ would then exist [AXIOM OF SUBCOLLECTIONS] contrary to the Zermelo/Russell paradox. Nevertheless it will occasionally occur that all the collections under discussion are subcollections of a particular 'universal' collection E (say); in this situation it will be convenient to adopt the convention that $\bigcap \varnothing = E$.

1.4 Days and histories

In this book we shall take the view that the creation of collections occurs step by step: in order to stress the temporal metaphor implicit in this, we shall say 'day by day'. The idea is that on the first day we start with the objects that already exist, i.e. the individuals; on each subsequent day we can collect together into a new collection any collections which were created on an earlier day. To simplify the development we identify a day with the collection of all collections in existence on that day. So the elements of each day will be precisely the individuals plus the elements and subcollections of all earlier days. The collection of all days earlier than a given day is called a 'history'. In fact the formal development proceeds more neatly if we define what we mean by a history first.

Definition

$\text{acc}(a) = \{x : x \text{ is an individual or } (\exists b \in a)(x \in b \text{ or } x \subseteq b)\}$ ('the *accumulation* of a').

Warning The axioms at our disposal do not yet permit us to prove that $\text{acc}(a)$ always exists.

Definition

$\mathscr{D}$ is called a *history* if

$$(\forall D \in \mathscr{D})(D = \text{acc}(\mathscr{D} \cap D)).$$

Example 1 Trivially $\varnothing$ is a history since it has no elements.

We turn now to proving theorem 1.4.2 below, which asserts (in the set-theoretic jargon) that the membership relation is well founded on each history. Although this result is central to exploiting the notion of a history its proof is somewhat technical and may be omitted on a first reading.

Definition

x is *grounded* if

$$(\forall a)(x \in a \Rightarrow (\exists y \in a)(y \cap a = \varnothing)).$$

In words (using terminology which will be discussed in §2.6): x is grounded if every collection to which x belongs has an $\in$-minimal element.

Remark 1 It will transpire later (once we have the axiom of creation at our disposal) that every object is grounded.

Definition

$\|b\| = \{x \in b : x \text{ is grounded}\}$ ('the *grounded part* of b').

Lemma 1.4.1

If $\mathscr{D}$ is a history and $D, D' \in \mathscr{D}$, then

$$D \in D' \Rightarrow \|D\| \in \|D'\|.$$

Proof Suppose on the contrary that $D \in D'$ but $\|D\| \notin \|D'\|$. Now $\|D\| \subseteq D$ and so $\|D\| \in D'$ by the definition of a history. Hence there exists $a \ni \|D\|$ such that $y \cap a \neq \varnothing$ for all $y \in a$ (since otherwise $\|D\| \in \|D'\|$). In particular, $\|D\| \cap a \neq \varnothing$, and therefore a has a grounded element. So $(\exists y \in a)(y \cap a = \varnothing)$. Contradiction. □

Theorem 1.4.2

If $\mathscr{D}$ is a history then

$$\varnothing \neq a \subseteq \mathscr{D} \Rightarrow (\exists D \in a)(D \cap a = \varnothing).$$

Proof Suppose that $D \in a \subseteq \mathscr{D}$ and let

$$b = \{\|D'\| : D' \in D \cap a\}.$$

Now if $b = \varnothing$ then $D \cap a = \varnothing$ and we are finished. So suppose from now on that $b \neq \varnothing$. Then since $b \subseteq \|D\|$ [lemma 1.4.1] there exists $D' \in D \cap a$ such that $\|D'\| \cap b = \varnothing$. We claim that $D' \cap a = \varnothing$. For if not then there exists $D'' \in D' \cap a$, so that $D'' \in D$ by the definition of a history, hence $\|D''\| \in \|D\|$ [lemma 1.4.1], so that $\|D''\| \in b$. But equally $\|D''\| \in \|D'\|$ [lemma 1.4.1] and so $\|D'\| \cap b \neq \varnothing$, which is absurd. □

Definition

The accumulation $\operatorname{acc}(\mathscr{D})$ of a history $\mathscr{D}$ (if it exists) is called the *day* with history $\mathscr{D}$.

Example 2 $\operatorname{acc}(\varnothing) = \{x : x \text{ is an individual}\}$, in the sense that if either side of the equation exists then they both do and they are equal.

Proposition 1.4.3

If D is a day with history $\mathscr{D}$ and $D' \in \mathscr{D}$ then D' is a day belonging to D with history $\mathscr{D} \cap D'$.

Proof Certainly $D' \subseteq D' \in \mathscr{D}$ and so $D' \in \text{acc}(\mathscr{D}) = D$. Also $D' = \text{acc}(\mathscr{D} \cap D')$ since $\mathscr{D}$ is a history. Moreover if $D'' \in \mathscr{D} \cap D'$ then $D'' \subseteq D'$, so that $\mathscr{D} \cap D'' = (\mathscr{D} \cap D') \cap D''$ and therefore

$$D'' = \text{acc}(\mathscr{D} \cap D'') = \text{acc}((\mathscr{D} \cap D') \cap D'').$$

Hence $\mathscr{D} \cap D'$ is a history. □

In the remainder of this chapter D, D', D_1, etc. will denote days.

In particular, the quantifiers $(\forall D)$ and $(\exists D)$ should be read 'For every day D' and 'For some day D' respectively.

Proposition 1.4.4

(*a*) $x \in b \in D \Rightarrow x \in D$.
(*b*) $a \in D \Rightarrow a \subseteq D$.
(*c*) $a \subseteq b \in D \Rightarrow a \in D$.
(*d*) $D = \{x : x \textit{ is an individual or } (\exists D' \in D)(x \subseteq D')\}$.

Proof Let $\mathscr{D}$ be a history of D.

(a) Suppose that $x \in b \in D$. If

$$a = \{D' \in \mathscr{D} : b \subseteq D' \text{ or } b \in D'\}$$

then $a \neq \varnothing$ by definition and so there exists $D' \in a$ such that $D' \cap a = \varnothing$ [theorem 1.4.2]. So either $b \subseteq D'$ or $b \in D'$. But if $b \in D'$ then (since b is not an individual) there exists $D'' \in \mathscr{D} \cap D'$ such that $b \in D''$ or $b \subseteq D''$ [proposition 1.4.3], and so $D'' \in D' \cap a$, contradicting the choice of D'. Hence $b \subseteq D'$, whence $x \in D' \in \mathscr{D}$ and so $x \in D$.

(b) Trivial from (a).

(c) Suppose that $a \subseteq b \in D$. As in the proof of part (a) we can obtain $D' \in \mathscr{D}$ such that $b \subseteq D'$. But then $a \subseteq D'$ and so $a \in D$.

(d) $x \in D \Leftrightarrow x$ is an individual or $(\exists D' \in \mathscr{D})(x \in D' \text{ or } x \subseteq D')$

$\Rightarrow x$ is an individual or $(\exists D' \in D)(x \in D' \text{ or } x \subseteq D')$ [proposition 1.4.3]

$\Leftrightarrow x$ is an individual or $(\exists D' \in D)(x \subseteq D')$ [(b)].

$\Rightarrow x \in D$ [(c)]. □

Remark 2 If $D_1 \in D_2$ we shall sometimes say that D_1 is *earlier than* D_2. With this terminology a day may be said to be the accumulation of all its yesterdays.

Remark 3 The hierarchy of days is *cumulative*: if an object belongs to a particular day, then it belongs to all subsequent days; put more emotively, objects are immortal.

Remark 4 We shall say that a collection c is *hereditary* if

$$a \subseteq b \in c \Rightarrow a \in c.$$

So proposition 1.4.4(c) asserts just that days are hereditary.

Lemma 1.4.5

If D is a day, then $\mathscr{D} = \{D' : D' \in D\}$ is a history whose day is D.

Proof Suppose that $D' \in D$. Then $D'' \in D' \Rightarrow D'' \in D$ [proposition 1.4.4(a)] and so $\mathscr{D} \cap D' = \{D'' : D'' \in D'\}$. Hence $D' = \operatorname{acc}(\mathscr{D} \cap D')$ since D' is a day. It follows that $\mathscr{D}$ is a history. □

Proposition 1.4.6

If Φ is a formula such that $(\exists D)\Phi(D)$, then

$$(\exists D_0)(\Phi(D_0) \text{ and not } (\exists D' \in D_0)\Phi(D')).$$

Proof Suppose that $\Phi(D)$ and let $a = \{D' \in D : \Phi(D')\}$. If $a = \varnothing$, then we can simply let $D_0 = D$. If not, then note that a is a subcollection of the history $\{D' \colon D' \in D\}$ [lemma 1.4.5] and so there exists $D_0 \in a$ such that $D_0 \cap a = \varnothing$ [theorem 1.4.2]; therefore $\Phi(D_0)$ and

$$D' \in D_0 \Rightarrow D' \in D \Rightarrow \text{not } \Phi(D'). \qquad \square$$

Proposition 1.4.7

$D_1 \in D_2$ *or* $D_1 = D_2$ *or* $D_2 \in D_1$.

Proof Suppose not. So we can find days D_1 and D_2 such that $D_1 \notin D_2$, $D_1 \neq D_2$, and $D_2 \notin D_1$: hence more particularly we can choose D_1 in such a way that

$$(\forall D \in D_1)(\forall D')(D \in D' \text{ or } D = D' \text{ or } D' \in D) \tag{1.4.1}$$

[Proposition 1.4.6] and, having chosen D_1, we can then choose D_2 in such a way that

$$(\forall D \in D_2)(D \in D_1 \text{ or } D = D_1 \text{ or } D_1 \in D) \tag{1.4.2}$$

[Proposition 1.4.6]. We shall now prove that with these choices of D_1 and D_2 we have

$$(\forall D)(D \in D_1 \Leftrightarrow D \in D_2). \tag{1.4.3}$$

Suppose first that $D \in D_1$. Then $D \neq D_2$ since $D_2 \notin D_1$. Also $D_2 \notin D$ since otherwise $D_2 \in D_1$ [proposition 1.4.4(a)], contrary to hypothesis. Since $D \in D_1$, we must therefore have $D \in D_2$ by (1.4.1). If $D \in D_2$, on the other hand, then the same arguments as before show that $D_1 \neq D$ and $D_1 \notin D$. Hence $D \in D_1$ by (1.4.2). This proves (1.4.3). But then

$$D_1 = \{x : x \text{ is an individual or } (\exists D \in D_1)(x \subseteq D)\}$$
[proposition 1.4.4(d)]
$$= \{x : x \text{ is an individual or } (\exists D \in D_2)(x \subseteq D)\}$$
[(1.4.3)]
$$= D_2 \quad \text{[proposition 1.4.4(d)].}$$
□

Definition

For any formula Φ such that $(\exists D)\Phi(D)$, we let

$$\mu D\Phi(D) = \mathbf{\iota}!D(\Phi(D) \text{ and not } (\exists D' \in D)\Phi(D'))$$

('the earliest D such that $\Phi(D)$').

Justification The existence of the term in question follows from Proposition 1.4.6 and its uniqueness is a consequence of proposition 1.4.7. □

Proposition 1.4.8

(*a*) $D \notin D$.
(*b*) $D \subseteq D' \Leftrightarrow (D \in D' \text{ or } D = D')$.
(*c*) $D \subseteq D'$ *or* $D' \subseteq D$.
(*d*) $D \subset D' \Leftrightarrow D \in D'$.

Proof

(a) If there exists a day D such that $D \in D$, then there exists an earliest such D: this leads immediately to a contradiction.
(b) If $D = D'$ then trivially $D \subseteq D'$; if $D \in D'$ then again $D \subseteq D'$ [proposition 1.4.4(b)]. If, on the other hand, neither $D \in D'$ nor $D = D'$ then $D' \in D$ [proposition 1.4.7] and $D' \notin D'$ [(a)], so that $D \nsubseteq D'$.
(c) If $D \nsubseteq D'$ then $D \notin D'$ and $D \neq D'$ [(b)], whence $D' \in D$ [proposition 1.4.7] and so $D' \subseteq D$ [(b)].
(d) $D \subset D' \Leftrightarrow (D \subseteq D' \text{ and } D \neq D') \Leftrightarrow D \in D'$ [(a) and (b)]. □

Exercise Show that these two assertions are equivalent:
(i) $(\forall a)(\exists D)(a \in D)$;
(ii) $(\forall a)(\exists D)(a \subseteq D)$ and $(\forall D)(\exists D')(D \in D')$.

1.5 Birthdays and the end of the world

The definitions we gave in the last section give us the vocabulary to describe a stage-by-stage process of collection creation. However, they do not guarantee that this process actually takes place. Indeed the axioms we have stated so far do not prevent $\varnothing$ from being the only collection.

So in order to rescue our theory from vacuity we shall from now on make the permissive assumption that the creation process can be carried out repeatedly: there is no last day. We shall also restrict our attention to the collections which occur somewhere in the iterative hierarchy. We achieve this by assuming that every collection is a subcollection of a day: more prosaically, every collection is well founded; less prosaically, every collection has a birthday. These two assumptions may be symbolized as follows:

(1) $(\forall D)(\exists D')(D \in D')$;

(2) $(\forall a)(\exists D)(a \subseteq D)$ (*The birthday principle*).

But to simplify matters we combine them into one axiom (cf. the exercise in §1.4).

Creation axiom

$(\forall a)(\exists D)(a \in D)$.

In words: every collection belongs to some day.

Remark 1 Our temporary axiom ensures that there is at least one collection (the empty one). The axiom of creation now ensures that there is at least one day (and hence that there are infinitely many).

Remark 2 We already know [proposition 1.4.4(d)] that every individual belongs to every day. Consequently the axiom of creation implies the apparently stronger

$$(\forall x)(\exists D)(x \in D);$$

every *object* belongs to some day.

Definition

The earliest day D such that $a \subseteq D$ is called the *birthday* of a and denoted $\mathrm{D}(a)$.

Justification Immediate [axiom of creation and proposition 1.4.4(b)].

□

Remark 3 We shall adopt the convention that the birthday of an individual is $\varnothing$. This is somewhat misleading since if there are any individuals $\varnothing$ is not a day. Nevertheless it accords with the account we have given of creation since it makes the individuals (if there are any) the objects of earliest possible birthday.

Proposition 1.5.1

$a \notin a$.

Proof If $a \in a$, then $a \in \mathrm{D}(a)$ (since $a \subseteq \mathrm{D}(a)$). Hence $(\exists D \in \mathrm{D}(a))(a \subseteq D)$ [proposition 1.4.4(d)], contradicting the definition of $\mathrm{D}(a)$. □

Definition

The earliest day of all is denoted D_0.

Proposition 1.5.2

$\mathrm{D}_0 = \{x : x \textit{ is an individual}\}$.

Proof Immediate [proposition 1.4.4(d)]. □

Proposition 1.5.3 (*Foundation principle*)

$a \subseteq \mathrm{D}_0$ *or* $(\exists b \in a)(b \cap a \subseteq \mathrm{D}_0)$.

Proof If $a \nsubseteq \mathrm{D}_0$, then we can choose a collection $b \in a$ of earliest possible birthday: if $y \in b \cap a$ then $y \in \mathrm{D}(b)$ and so if y is not an individual there exists $D' \in \mathrm{D}(b)$ such that $y \subseteq D'$ [proposition 1.4.4(d)], i.e. $\mathrm{D}(y)$ is earlier than $\mathrm{D}(x)$, which contradicts the fact that $y \in a$; therefore $b \cap a \subseteq \mathrm{D}_0$ [proposition 1.5.2] as required. □

Remark 4 If we had assumed that individuals have no elements (cf. remark 1 of §1.2), we could have stated the foundation principle in the form

$$a \neq \varnothing \Rightarrow (\exists x \in a)(x \text{ is an individual or } x \cap a = \varnothing).$$

Well foundedness By assuming that every collection has a birthday we make the cumulative, iterative, well-founded hierarchy of days central to our conception of the theory. Mirimanoff (1917) was the first to focus attention on the hierarchy but it was Scott (1974) who had the idea of making it fundamental to an axiomatic presentation and we have followed his treatment here. However, Scott took the notion of 'day' as primitive and assumed an extra axiom—the 'axiom of accumulation'—which asserted that (in our terminology) every day accumulates all its yesterdays; it was Derrick who showed how to *define* days and eliminate Scott's axiom. (Derrick's unpublished work analyses the structure of the binary predicates which can be shown in our system to be well founded.)

All this is in contrast to the historical development of axiomatizations, in which a corresponding assumption (essentially what we have called the foundation principle) did not appear until von Neumann (1925) and Zermelo (1930). Some treatments (notably Bourbaki 1954) still omit such an axiom altogether. The consequences of doing this are not great: the definitions of cardinal and ordinal numbers which we shall give in Chapters 4 and 5 do not apply to collections which are not well founded, but we could easily postulate what we want instead. However, the metamathematical study of set theory which has been carried out in the last 50 years has exploited well foundedness systematically. It is no doubt rather dogmatic to ban collections which are not well founded. Our reason for doing so is a desire for simplicity, not a belief that they are unworthy of a mathematician's attention. Forms of set theory which not only allow but actively encourage non-well-foundedness have arisen recent, principally in the theory of concurrent processes (Aczel 1988).

Predicative set theory The AXIOM OF SUBCOLLECTIONS appears to violate the temporal account we have given of the creation of collections because it allows the possibility that the formula $\Phi(x)$ which defines the collection $\{x \in a : \Phi(x)\}$ may depend on variables representing collections which are created *after* a: definitions of this kind are called *impredicative*. Of course there is nothing wrong with impredicative definitions in themselves: 'the tallest man in the world' is impredicative but unobjectionable. The worry occurs because of the explicitly hierarchical (temporal) account of collection creation which we have given. On the strong realist view which allows there to be indefinable collections, this is not really a problem: when we create $\mathfrak{P}(D)$ we collect together all subcollections of D in as strong a sense of 'all' as possible (the maximal conception of the formation of power collections); hence the AXIOM OF SUBCOLLECTIONS asserts much less than is the case, not more.

The alternative view, whereby we can only create at each stage those collections which can be defined in terms of collections which have already been created, would lead us to replace the AXIOM OF SUBCOLLECTIONS by a *predicative* version such as the following: if $\Phi(x)$ is a formula and $x_1, \ldots, x_n$ are the variables other than x on which it depends, then

$$(\forall x_1, \ldots, x_n \in D)(\{x \in D : \Phi^{(D)}(x)\} \text{ exists})$$

is an axiom*. Such a predicative form of the AXIOM OF SUBCOLLECTIONS would restrict quite substantially what we could prove in our theory.

Definition

We introduce the following abbreviations:

$$\bigcup a = \{x : (\exists b \in a)(x \in b)\} \quad \text{('the } \textit{union} \text{ of } a\text{')}$$

$$\mathfrak{P}(a) = \{b : b \subseteq a\} \quad \text{('the } \textit{power} \text{ of } a\text{')}$$

$$a \cup b = \{x : x \in a \text{ or } x \in b\}.$$

* For the definition of $\Phi^{(D)}$ see p. 58.

Justification If we start with a collection a, then there exists D such that $a \in D$ [creation axiom]. Now

$$(b \in D \text{ and } b \subseteq a) \Leftrightarrow b \subseteq a$$

[proposition 1.4.4(b)], whence $\mathfrak{P}(a) = \{b \in D : b \subseteq a\}$ exists [AXIOM OF SUBCOLLECTIONS]. Also

$$x \in b \in a \in D \Rightarrow x \in b \in D \Rightarrow x \in D$$

[proposition 1.4.4(a)]; hence

$$\bigcup a = \{x \in D : (\exists b \in a)(x \in b)\}$$

exists [AXIOM OF SUBCOLLECTIONS]. Suppose now that we have a second collection b, so that there exists D' such that $b \in D'$ [creation axiom]. Now either $D \subseteq D'$ or $D' \subseteq D$ [proposition 1.4.8(c)]: suppose the latter for the sake of argument. Then $a, b \in D$ and so

$$(x \in a \text{ or } x \in b) \Rightarrow x \in D$$

[proposition 1.4.4(b)]. Hence

$$a \cup b = \{x \in D : x \in a \text{ or } x \in b\}$$

exists [AXIOM OF SUBCOLLECTIONS]. □

Proposition 1.5.3

(*a*) $a \in D \Rightarrow \bigcup a \in D$.
(*b*) $a, b \in D \Rightarrow a \cup b \in D$.

Proof (a) If $a \in D$, then $a \subseteq D'$ for some $D' \in D$ [proposition 1.4.4(d)]. Consequently

$$\begin{aligned} x \in \bigcup a &\Rightarrow x \in b \in a \quad \text{for some } b \\ &\Rightarrow x \in b \in D' \\ &\Rightarrow x \in D' \quad \text{[proposition 1.4.4(b)].} \end{aligned}$$

So $\bigcup a \subseteq D'$ and therefore $\bigcup a \in D$ [proposition 1.4.4(d)].
(b) Exercise. □

Definition

$\{x\} = \{y : y = x\}$ ('the *singleton* of x').

Justification There is a day D such that $x \in D$ by remark 2. So $\{x\} = \{y \in D : x = y\}$, which exists [AXIOM OF SUBCOLLECTIONS]. □

Definition

$\{x, y\} = \{x\} \cup \{y\}$, $\{x, y, z\} = \{x, y\} \cup \{z\}$, etc.

Remark 5 We have already remarked on the significance of the distinction between the object x and the singleton $\{x\}$. However, it is a common practice among mathematicians, which we shall occasionally follow, to denote $\{x\}$ simply by x when it occurs in compound expressions. Put more formally, if symb(a) is an expression symbolizing a term depending on the variable a, then we write symb(x) for symb($\{x\}$) if no confusion is likely to arise thereby. Similarly, we may sometimes write symb(x, y) for symb($\{x, y\}$), etc.

Definition

If $\tau(x_1,\ldots,x_n)$ is a term depending on the variables $x_1, \ldots, x_n$ and $\Phi(x_1,\ldots,x_n)$ is a formula, then we write $\{\tau(x_1,\ldots,x_n):\Phi(x_1,\ldots,x_n)\}$ as an abbreviation for

$$\{y:(\exists x_1,\ldots,x_n)(y=\tau(x_1,\ldots,x_n) \text{ and } \Phi(x_1,\ldots,x_n))\}.$$

In the particular case when $\tau(x) = x$ this coincides with our previous notation.

Remark 6 If we also have at our disposal an expression symb(a) (say) to denote a term depending on a, then the term denoted by the expression symb($\{\tau(x_1,\ldots,x_n):\Phi(x_1,\ldots,x_n)\}$) may also be denoted $\text{symb}_{\Phi(x_1,\ldots,x_n)}\,\tau(x_1,\ldots,x_n)$.

Example $\bigcup_{x\in a}\tau(x)$ [respectively $\bigcap_{x\in a}\tau(x)$] is another way of writing $\bigcup\{\tau(x):x\in a\}$ [respectively $\bigcap\{\tau(x):x\in a\}$].

Remark 7 Of course, the term τ and the formula Φ may depend on other variables apart from $x_1, \ldots, x_n$. If so, then the notations we have just introduced are, strictly speaking, ambiguous, since it may be unclear which are the 'active' variables. In practice, however, this is never a problem.

Exercises

1. (a) Show that $a \subseteq b \Rightarrow \mathrm{D}(a) \subseteq \mathrm{D}(b)$.
   (b) Show that $a \in b \Rightarrow \mathrm{D}(a) \in \mathrm{D}(b)$.
2. (a) If Φ is a formula, show that

$$(\exists a)\Phi(a) \Rightarrow (\exists a)(\Phi(a) \text{ and not } (\exists b\in a)\Phi(b)).$$

   [Consider the earliest D such that $(\exists a\in D)\Phi(a)$.]
   (b) Show that there do not exist a and b such that $a \in b$ and $b \in a$.
3. Show that $x = y \Leftrightarrow (\forall a)(x\in a \Rightarrow y \in a)$.
4. Prove proposition 1.5.3(b).
5. Show that $D = \mathrm{D}_0 \cup \bigcup_{D'\in D}\mathfrak{P}(D')$.

We mention now a technical device which is sometimes useful when we want to use a collection $\{x : \Phi(x)\}$ which does not exist. The idea is that instead of attempting the doomed task of forming the non-existent collection of *all* those x such that $\Phi(x)$, we restrict ourselves to collecting only such x of earliest possible birthday, thereby creating a collection which always exists and which may retain enough information about Φ to be useful.

Definition

If $\Phi(x)$ is a formula then

$$\langle x : \Phi(x)\rangle = \{x : \Phi(x) \text{ and } (\forall y)(\Phi(y) \Rightarrow \mathrm{D}(x) \subseteq \mathrm{D}(y))\}.$$

Justification Straightforward. □

Proposition 1.5.4

If $\Phi(x)$ is a formula, then:

$$(\exists y \in D)\Phi(y) \Leftrightarrow \varnothing \neq \langle x : \Phi(x)\rangle \subseteq D.$$

Proof Suppose first that $y \in D$ and $\Phi(y)$. Then there exist elements z of earliest possible birthday such that $\Phi(z)$, i.e. $\langle x : \Phi(x)\rangle \neq \varnothing$. Moreover, if $z \in \langle x : \Phi(x)\rangle$ then either z is an individual, in which case certainly $z \in D$, or it is a collection, in which case $z \subseteq \mathrm{D}(z) \subseteq \mathrm{D}(y) \in D$ and hence again $z \in D$; thus $\langle x : \Phi(x)\rangle \subseteq D$. The converse implication is trivial. □

Exercise 6. What is $\langle x : x \notin x\rangle$? (Recall that $\{x : x \notin x\}$, on the other hand, does not exist [Zermelo/Russell paradox].)

Definition

$\mathrm{cut}(a) = \langle x : x \in a\rangle$ ('the *cutting-down* of a').

Corollary 1.5.5

$a \cap D \neq \varnothing \Rightarrow \varnothing \neq \mathrm{cut}(a) \subseteq D.$

Proof Immediate [Proposition 1.5.4]. □

We are now in a position to clarify the way in which days are created. First some terminology: the *day after* D is the earliest day D' such that $D \in D'$; a day D is called a *limit day* if it is neither D_0 nor the day after any other day.

Proposition 1.5.6

The day after D is $\mathrm{D}_0 \cup \mathfrak{P}(D)$.

Proof Let D' be the day after D.

Now $D'' \subseteq D \Rightarrow D'' \in D'$ [proposition 1.4.4(c)] and

$$\begin{aligned} D'' \in D' &\Rightarrow D \notin D'' \\ &\Rightarrow (D'' \in D \text{ or } D'' = D) \quad \text{[proposition 1.4.7]} \\ &\Rightarrow D'' \subseteq D \quad \text{[proposition 1.4.8(b)].} \end{aligned}$$

Hence

$$D'' \in D' \Leftrightarrow D'' \subseteq D. \tag{1.5.1}$$

So

$$\begin{aligned} D' &= \{x : x \text{ is an individual or } (\exists D'' \in D')(x \subseteq D'')\} \text{ [proposition 1.4.4(d)]} \\ &= \{x : x \text{ is an individual or } (\exists D'' \subseteq D)(x \subseteq D'')\} \quad [(1.5.1)] \\ &= \{x : x \text{ is an individual or } x \subseteq D\} \\ &= \mathrm{D}_0 \cup \mathfrak{P}(D) \quad \text{[proposition 1.5.2].} \end{aligned}$$

□

Remark 8 As a consequence of propositions 1.5.2 and 1.5.6 we can see that the creation process begins with the days D_0, $\mathrm{D}_0 \cup \mathfrak{P}(\mathrm{D}_0)$, $\mathrm{D}_0 \cup \mathfrak{P}(\mathfrak{P}(\mathrm{D}_0))$, $\mathrm{D}_0 \cup \mathfrak{P}(\mathfrak{P}(\mathfrak{P}(\mathrm{D}_0)))$, etc. Indeed, not only is no day the last day of creation but after each day there are infinitely many still to come, which is comforting.

Proposition 1.5.7

If D is a day other than D_0, then these five assertions are equivalent:

(*i*) D *is a limit;*
(*ii*) $(\forall a \in D)(\exists D' \in D)(a \in D')$;
(*iii*) $a \in D \Rightarrow \mathfrak{P}(a) \in D$;
(*iv*) $a \in D \Rightarrow \{a\} \in D$;
(*v*) $a \subseteq D \Rightarrow \mathrm{cut}(a) \in D$.

(i) ⇒ (ii) Suppose that D is a limit day and $a \in D$. So there exists $D' \in D$ such that $a \subseteq D'$, hence $a \in \mathrm{D}_0 \cup \mathfrak{P}(D')$. But $\mathrm{D}_0 \cup \mathfrak{P}(D')$ is the day after D' [proposition 1.5.6] and therefore $\mathrm{D}_0 \cup \mathfrak{P}(D') \subset D$ since D is a limit, i.e. $\mathrm{D}_0 \cup \mathfrak{P}(D') \in D$.
(ii) ⇒ (iii) Suppose $a \in D$. Then there exists $D' \in D$ such that $a \in D'$. So $x \subseteq a \Rightarrow x \in D'$ [proposition 1.4.4(c)]. Therefore $\mathfrak{P}(a) \subseteq D'$ and hence $\mathfrak{P}(a) \in D$.
(iii) ⇒ (iv) If $a \in D$ then by hypothesis $\{a\} \subseteq \mathfrak{P}(a) \in D$ and therefore $\{a\} \in D$ [proposition 1.4.4(c)].
(iv) ⇒ (i) If D is not a limit, the there exists $D' \in D$ such that $D = \mathrm{D}_0 \cup \mathfrak{P}(D')$ [proposition 1.5.6]. By hypothesis $\{D'\} \in D$, i.e. $\{D'\} \subseteq D'$, so that $D' \in D'$, which contradicts proposition 1.4.7(a).

(ii) $\Rightarrow$ (v) If $a = \varnothing$ then $\mathrm{cut}(a) = \varnothing$ and the result is trivial. If not, then there exist an element $x \in a$ and therefore by hypothesis a day $D' \in D$ such that $x \in D'$: hence $\mathrm{cut}(a) \subseteq D'$ and so $\mathrm{cut}(a) \in D$ [proposition 1.4.4(c)].
(v) $\Rightarrow$ (iv) Suppose that $a \in D$ and let $b = \{c : a \subseteq c \in D\}$. Then $b \subseteq D$ and so by hypothesis $\mathrm{cut}(b) \in D$. But it is easy to verify that $\{a\} \subseteq \mathrm{cut}(b)$, and so $\{a\} \in D$ [proposition 1.4.4(c)]. □

The theory whose axioms we have stated in this chapter—the AXIOM OF SUBCOLLECTIONS, the axiom of creation, and the temporary axiom—is called elementary collection theory.

2 Relations

Now that we have an axiomatic base for the theory of collections in place we are in a position to define the familiar apparatus of pure mathematics: functions, equivalence relations, orderings, etc.

2.1 Ordered pairs and relations

First we show how to define a pairing term in our theory.

Definition

$(x, y) = \{\{x\}, \{x, y\}\}$ ('the *ordered pair* of x and y').

Lemma 2.1.1

$\{x, y\} = \{x, z\} \Rightarrow y = z.$

Proof $y \in \{x, y\} = \{x, z\}$, so that either $y = z$ as required or $y = x$: but if $y = x$, then $z \in \{x, z\} = \{x, y\} = \{y\}$, and so $z = y$ in this case as well. □

Proposition 2.1.2

(x, y) is a pairing term.

Proof Suppose that $(x, y) = (z, t)$, i.e. $\{\{x\}, \{x, y\}\} = \{\{z\}, \{z, t\}\}$. So either $\{x\} = \{z\}$, in which case $x = z$, or $\{x\} = \{z, t\}$, in which case $x = z = t$. Hence in either case $\{x\} = \{z\}$, so that $\{x, y\} = \{z, t\}$ [lemma 2.1.1] and therefore $y = t$ [lemma 2.1.1]. □

Ordered pairs The first satisfactory definition of ordered pairs—essentially that of exercise 1 below—was given by Wiener (1914); a slightly less satisfactory one appears in Hausdorff (1914). Kuratowski's definition—the one we have just adopted—arose out of a characterization he gave in 1921 of total orderings in terms of their initial subcollections and was part of his work on extending the scope of Zermelo's axiomatization: Zermelo's system did not have the notion of an ordered pair at all. The technical advantage of being able to define a pairing term is clear: Whitehead and Russell (1910–13) were forced by ignorance of such devices to develop two parallel but distinct theories, one of classes and one of relations. Kuratowski's definition, in particular, is nowadays widely enough used that it is almost universal. Nevertheless, all the explicit definitions of pairing terms have, quite apart from their arbitrariness, the disadvantage that they do too much: Kuratowski's definition, for example, has the entirely accidental consequence contained in exercise 2(a) below. For this reason some authors (e.g. Bourbaki prior to the 5th edition of *Théorie des ensembles*) have opted to regard the expression (x, y) as a primitive term and proposition 2.1.2 as an axiom.

Exercise 1. Show that $\{\{\{x\}, \varnothing\}, \{\{y\}\}\}$ is a pairing term.

Definition

If z is an ordered pair, let:

$$\mathrm{dom}(z) = \iota! x(\exists y)(z = (x, y)) \quad \text{('the \textit{first coordinate} of } z\text{');}$$

$$\mathrm{im}(z) = \iota! y(\exists x)(z = (x, y)) \quad \text{('the \textit{second coordinate} of } z\text{').}$$

Justification Immediate [proposition 2.1.2]. □

Remark 1 $\mathrm{dom}(x, y) = x$ and $\mathrm{im}(x, y) = y$.

Definition

$A \times B = \{z : \mathrm{dom}(z) \in A \text{ and } \mathrm{im}(z) \in B\}$ ('the *cartesian product* of A and B').

In words rather than symbols: $A \times B$ is the collection of ordered pairs whose first coordinate is in A and whose second coordinate is in B.

Proposition 2.1.3

Suppose that D is a limit day.
(a) $a, b \in D \Rightarrow (a, b) \in D$.
(b) $A, B \in D \Rightarrow A \times B \in D$.

Proof This follows at once from proposition 1.5.7 since $(a, b) \subseteq \mathfrak{P}(\{a, b\})$ and $A \times B \subseteq \mathfrak{P}(\mathfrak{P}(A \cup B))$. □

Remark 2 The birthday of $\{a, b\}$ is the day after the later of the birthdays of a and b; the birthday of (a, b) is the day after that.

Exercise 2. (a) Show that $\{a\} \times \{a\} = \{\{\{a\}\}\}$.
(b) Show that $A \times A = A \Leftrightarrow A = \varnothing$. [If A is non-empty, consider an element of A of earliest possible birthday.]
(c) Give an example of collections X, Y, Z, T such that $X \times Y = Z \times T$ but $X \neq Z$ and $Y \neq T$.

Definition

A collection is called a *relation* if every element of it is an ordered pair; a relation which is a subcollection of $A \times B$ is said to be *between* A and B.

If $\Phi(x, y)$ is a formula, then $\{z \in A \times B : \Phi(\mathrm{dom}(z), \mathrm{im}(z))\}$ (which is usually written $\{(x, y) \in A \times B : \Phi(x, y)\}$ of course) is a relation between A and B: it is said to be the relation between x in A and y in B *defined by* the formula $\Phi(x, y)$. Conversely, if r is a relation between A and B, then the formula $(x, y) \in r$, which is customarily written $x\ r\ y$, defines the given relation r.

If r is a relation between A and B, then the collection $\{(y, x) \in B \times A : x\ r\ y\}$ is called the *inverse* of r and denoted r^{-1}. The collections

$$\mathrm{dom}[r] = \{\mathrm{dom}(z) : z \in r\}$$

$$\mathrm{im}[r] = \{\mathrm{im}(z) : z \in r\}$$

are called the *domain* and *image* of r respectively. If r and s are relations, then we let $r \circ s$ denote the relation

$$\{(x, z) : (\exists y)(x\ s\ y \text{ and } y\ r\ z)\}.$$

Warning The order of r and s in the definition of $r \circ s$ is not what one might expect. (See remark 2 of §2.2.)

Suppose that r is a relation between A and B. If $C \subseteq A$, then $r \cap (C \times B)$ is called the *restriction* of r to C and denoted $r|C$; if $r|C = s|C$, we say that r and s *agree on* C. We let $r[C]$ denote the collection

$$\mathrm{im}[r|C] = \{y : (\exists x \in C)(x\ r\ y)\}.$$

Exercise

3. Suppose that r, s, and t are relations.
   (a) Show that $(r \circ s) \circ t = r \circ (s \circ t)$.
   (b) Show that $(r \circ s)^{-1} = s^{-1} \circ r^{-1}$.

2.2 Functions and mappings

A relation f between A and B is said to be *functional* if for every $x \in A$ there is exactly one $y \in B$ such that $x\ f\ y$. Functional relations between A and B are more usually called *functions from A to B*, or *families in B indexed by A*. The collection of all such functions is denoted ${}^{A}B$. If $\tau(x)$ is a term which denotes an element of B for every $x \in A$, then the collection $\{(x, \tau(x)) : x \in A\}$ is a function from A to B which is said to be *defined* by the term $\tau(x)$; this function is denoted $(\tau(x))_{x \in A}$ or '$x \mapsto \tau(x)$ $(x \in A)$'. If the domain is clear from the context, we denote the function simply by

$(\tau(x))$ or '$x \mapsto \tau(x)$'. Conversely, if f is a function from A to B then $\iota!y(x f y)$ exists for every $x \in A$; this term defines the relation f and is denoted $f(x)$. If f and g are functions then $(f \circ g)(x) = f(g(x))$ for all x.

Remark 1 Of course, the term $(\tau(x))_{x \in A}$ does not depend on x; in this expression, the letter x is being used as a dummy variable and could be replaced by any other letter which does not already occur in the expression.

Remark 2 The fact that the value of the function f at x is almost universally denoted $f(x)$ rather than $(x)f$ or $x|f$ is an historical accident with nothing except tradition to commend it*.

Remark 3 The languages of functions and families have different connotations and it is not usual to mix them. Nevertheless their mathematical *meanings* are identical. The following dictionary will enable you to translate from one to the other:

Function f	Family $(a_x)_{x \in A}$
Domain A	Indexing collection A
Image $f[A]$	Range $\{a_x : x \in A\}$
Value $f(x)$	Term a_x

Lemma 2.2.1

If $(f_i)_{i \in I}$ is a family of functions such that $f_i \cup f_j$ is a function for all $i, j \in I$, then $\bigcup_{i \in I} f_i$ is a function.

Proof Suppose that $(x, y_1), (x, y_2) \in \bigcup_{i \in I} f_i$. So there exist $i_1, i_2 \in I$ such that $(x, y_1) \in f_{i_1}$ and $(x, y_2) \in f_{i_2}$. So (x, y_1) and (x, y_2) belong to $f_{i_1} \cup f_{i_2}$, which is a function by hypothesis, and therefore $y_1 = y_2$. □

Suppose that f is a function from A to B. The ordered pair $(A \times B, f)$ is called the *mapping $A \to B$ determined by f* and denoted $f : A \to B$. The collections f, A, and B are called respectively the *graph*, *domain*, and *codomain* of the mapping $f : A \to B$. The mapping $f : A \to B$ is said to be *injective* [respectively *surjective*, *bijective*] if for each $y \in B$ there exists at most [respectively at least, exactly] one $x \in A$ such that $y = f(x)$. The graph of an injective mapping is said to be *one-to-one*; the graph of a bijective mapping $A \to B$ is sometimes called a *one-to-one correspondence*

* Dedekind used the notation $x|f$ in an early draft of *Was sind und was sollen die Zahlen?* (Dugac 1976, appendice LVI); but he used $f(x)$ in the published version.

between A and B. If there exists a one-to-one correspondence between A and B, then we say that A and B are *equinumerous.*

Example 1 If $B \subseteq A$ then the mapping $\mathrm{id}_B : B \to A$ given by $x \mapsto x$ $(x \in B)$ is injective: it is called the *canonical inclusion mapping* of B into A. The graph of this mapping is the collection $\mathrm{id}_B = \{(x, x) : x \in B\}$, which is sometimes called the *diagonal* of B or the *identity function* on B. (It might be more logical to write $=_B$ instead of id_B but there are contexts in which this looks very odd.)

Example 2 The mapping $\mathrm{id}_A : A \to A$ is bijective: it is called the *identity mapping* on A.

Remark 4 Evidently two mappings are equal iff they have the same graph, domain, and codomain. Of course, the explicit presence of the domain in the formalism for a mapping is redundant since it is determined by the graph.

Remark 5 The distinction between the barred arrow $\mapsto$ and the straight arrow $\to$ is a useful one: the former tells us how the mapping acts on an element $(x \mapsto f(x))$; the latter tells us what the domain and codomain are $(A \to B)$.

Remark 6 Situations involving several mappings are often represented by *mapping diagrams.* For example, the diagram

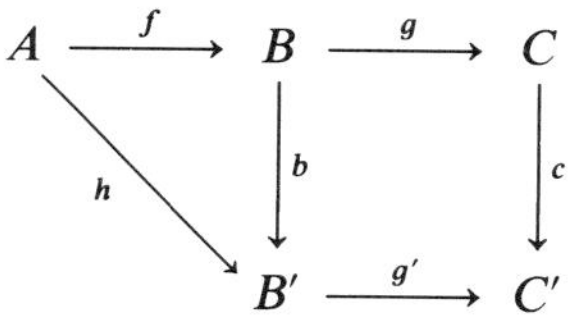

represents a situation in which there are mappings $f : A \to B$, $g : B \to C$, $g' : B' \to C'$, $h : A \to B'$, $b : B \to B'$, and $c : C \to C'$. A mapping diagram can be thought of as an oriented graph in which the nodes are the domains and codomains (in this case A, B, C, B', and C') and the directed edges are the functions (in this case f, g, g', h, b, and c). In addition to the functions explicitly displayed, there is for each collection in the diagram an identity function which can be thought of as a path from that node to itself: we shall adopt the convention that these identity functions are implicit in the diagram. A mapping diagram is said to *commute* if following two different routes between the same nodes and composing the functions concerned produces the same result. To assert that the above diagram commutes is therefore tantamount to asserting that $c \circ g = g' \circ b$ and $b \circ f = h$.

The following characterizations of injective and surjective mappings state the obvious in a rather perverse way. However, we state them here because they serve as prototypes for the definition of the analogous concepts in a great variety of situations.

Proposition 2.2.2

A mapping $f : A \to B$ is injective iff for every mapping $g : C \to B$ such that $\mathrm{im}[g] \subseteq \mathrm{im}[f]$ *there is a unique mapping $h : C \to A$ which makes the diagram*

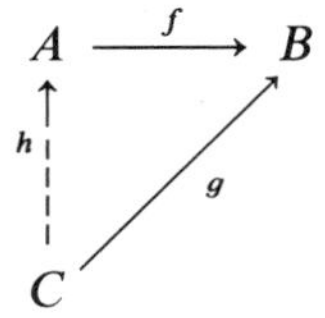

commute.

Necessity In the circumstances described, for each $x \in C$ there is exactly one $y \in A$ such that $f(y) = g(x)$. So if we define

$$h(x) = \iota ! y(f(y) = g(x))$$

we get the unique function with the properties we require.

Sufficiency Apply the condition in the special case when $C = A$ and $g = f$. Then letting h be the identity function on A suffices to make the diagram commute. Suppose now that $f(x) = f(y)$: then we could equally let h be the function which swaps x and y but leaves every other element of A fixed; since h is unique, it follows that $x = y$. Thus $f : A \to B$ is injective. □

Corollary 2.2.3

If $A \subseteq B$ then for every mapping $g : C \to B$ such that $\mathrm{im}[g] \subseteq A$ *the mapping $g : C \to A$ is the unique mapping which makes the diagram*

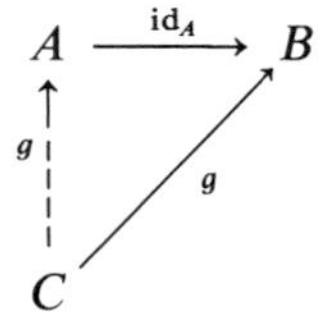

commute.

Proof Immediate [proposition 2.2.2]. □

Definition

If $f : B \to A$ is a mapping, then the collection

$$\{(x, y) \in B \times B : f(x) = f(y)\}$$

is called the *equivalence kernel* of f and denoted $\mathrm{Ker}[f]$.

Proposition 2.2.4

A mapping $f : B \to A$ is surjective iff for every mapping $g : B \to C$ such that $\mathrm{Ker}[f] \subseteq \mathrm{Ker}[g]$ *there is a unique mapping $h : A \to C$ which makes the diagram*

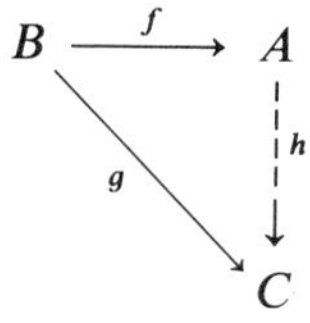

commute.

Necessity In the circumstances described in the proposition, for each $x \in A$ there is exactly one $y \in C$ such that $y = g(z)$ for some $z \in B$ such that $f(z) = x$. Consequently, if we define

$$h(x) = \iota!y(\exists z)(f(z) = x \text{ and } g(z) = y),$$

we get the unique function with the properties we require.

Sufficiency Apply the condition in the special case when $C = A \cup \{\infty\}$ (where ∞ is some object which does not belong to A) and $g = f$. Then letting h be the identity function on A obviously suffices to ensure that the diagram commutes. But if there exists $x \in A - f[B]$ then we could equally send x to ∞ and leave every other element of A fixed; since h is unique, such an x cannot exist. So $f : B \to A$ is surjective. □

Remark 7 Note the duality between propositions 2.2.2 and 2.2.4: to get from one to the other reverse the arrows.

The definition we gave in §2.1 of the cartesian product $A \times B$ of two collections may evidently be extended to three, four, or more collections. But it is not directly generalizable to a definition of the product of an arbitrary family $(A_i)_{i \in I}$ of collections. To achieve this we must use a rather different approach.

Definition

If $(A_i)_{i \in I}$ is a family of collections, then the collection

$$\left\{ f \in {}^{I}\!\left(\bigcup_{i \in I} A_i \right) : (\forall i \in I)(f(i) \in A_i) \right\}$$

is called the *product* of the family and denoted $\Pi_{i \in I} A_i$. For each $j \in I$ the jth *projection* is the mapping $\mathrm{pr}_j : \Pi_{i \in I} A_i \to A_j$ given by $\mathrm{pr}_j(f) = f(j)$.

Expressed in the language of families rather than of functions, $\Pi_{i \in I} A_i$ consists of the families $(x_i)_{i \in I}$ such that $x_i \in A_i$ for all $i \in I$.

Example 3 If $A_i = A$ for all $i \in I$ then $\Pi_{i \in I} A_i = {}^I A$.

Proposition 2.2.5

If B is a collection, $(A_i)_{i \in I}$ is a family of collections, and $f_i : B \to A_i$ is a mapping for each $i \in I$, then there exists a unique mapping $f : B \to \Pi_{i \in I} A_i$ which makes the diagram

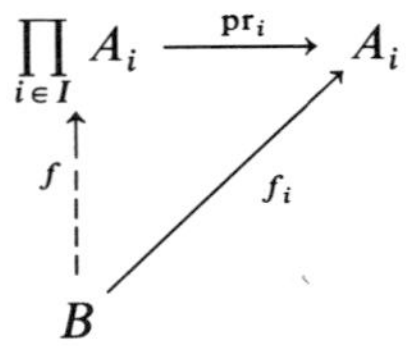

commute for all $i \in I$.

Proof It is clear that we must let $f(x) = (f_i(x))_{i \in I}$ for all $x \in B$. □

Remark 8 Let us choose (temporarily) two collections, which we shall denote 0 and 1, and let $\mathbf{2} = \{0, 1\}$. (This is a particular case of a notation we shall introduce in §4.3.) If A_0 and A_1 are collections then there is evidently a definite bijective mapping $\Pi_{r \in \mathbf{2}} A_r \to A_0 \times A_1$ given by $(x_r)_{r \in \mathbf{2}} \mapsto (x_0, x_1)$. It is usual to identify $\Pi_{r \in \mathbf{2}} A_r$ with $A_0 \times A_1$ by means of this mapping and to regard the cartesian product as a special case of the general product.

Each element of the union $\bigcup_{i \in I} A_i$ belongs to A_i for some index i. But which one? If the A_i are not disjoint there may be more than one such index. It is sometimes useful to construct a union in such a way that it keeps the A_i disjoint.

Definition

If $(A_i)_{i \in I}$ is a family of collections, then the collection $\bigcup_{i \in I} (A_i \times \{i\})$ is called the *disjoint union* of the family $(A_i)_{i \in I}$ and denoted $\dot{\bigcup}_{i \in I} A_i$. For each $j \in I$ the jth *insertion* is the mapping $\mathrm{p}_j : A_j \to \dot{\bigcup}_{i \in I} A_i$ given by $\mathrm{p}_j(x) = (x, j)$.

Informally: to form the union of a family of sacks, empty the contents of all of them into one big sack; to form the disjoint union, tag each object beforehand with a label saying which sack it was in.

Example 4 If $A_i = A$ for all $i \in I$ then $\dot{\bigcup}_{i\in I} A_i = A \times I$.

Proposition 2.2.6

If B is a collection, $(A_i)_{i\in I}$ is a family of collections, and $f_i : A_i \to B$ is a mapping for each $i \in I$, then there exists a unique mapping $f : \dot{\bigcup}_{i\in I} A_i \to B$ which makes the diagram

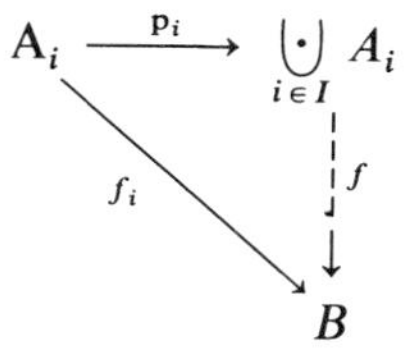

commute for all $i \in I$.

Proof We must evidently let $f(x, i) = f_i(x)$ for all $i \in I$ and $x \in A_i$. □

Remark 9 We have here another example of arrow-reversing duality, this time between the mapping properties for products and disjoint unions (propositions 2.2.5 and 2.2.6).

Remark 10 There are canonical mappings $\mathrm{dom} : \dot{\bigcup}_{i\in I} A_i \to \bigcup_{i\in I} A_i$ given by $(x, i) \mapsto x$ and $\mathrm{im} : \dot{\bigcup}_{i\in I} A_i \to I$ given by $(x, i) \mapsto i$. Note that the latter mapping remembers the index of the collection A_i which an element of $\dot{\bigcup}_{i\in I} A_i$ belongs to. Note also that the former mapping is surjective (and is bijective iff the A_i are disjoint) whereas the latter mapping is surjective iff the A_i are all non-empty.

Remark 11 In the notation of remark 8 above we write $A_0 \dot{\cup} A_1$ to denote $\dot{\bigcup}_{r\in \mathbf{2}} A_r$. In other words $A_0 \dot{\cup} A_1 = (A_0 \times \{0\}) \cup (A_1 \times \{1\})$. (Note that $A \dot{\cup} A = A \times \mathbf{2}$.)

Exercises

1. If $f : A \to B$ and $g : B \to A$ are mappings such that $g \circ f = \mathrm{id}_A$ then we say that g is the *left inverse* of f and f is the *right inverse* of g.
   (a) Show that a mapping $f : A \to B$ is injective iff either $A = \varnothing$ or f has a left inverse.
   (b) Show that if $g : B \to A$ has a right inverse, then it is surjective. (For the converse see exercise 3 of §7.3.)
   (c) Show that $f : A \to B$ is bijective iff it has both a left inverse and a right inverse.
2. Suppose that $(A_i)_{i\in I}$ and $(B_i)_{i\in I}$ are families of collections and that $\prod_{i\in I} A_i \neq \varnothing$.
   (a) Show that $\prod_{i\in I} A_i \subseteq \prod_{i\in I} B_i \Leftrightarrow (\forall i \in I)(A_i \subseteq B_i)$.
   (b) Show that $\mathrm{pr}_j[\prod_{i\in I} A_i] = A_j$ for all $j \in I$.

2.3 Relations on a collection

Definition

A relation between a collection A and itself (i.e. a subcollection of $A \times A$) is said to be a relation *on* A.

Definition

If r is a relation on A, then a collection $B \subseteq A$ is said to be *r-closed* if $r[B] \subseteq B$. The intersection of all the r-closed subcollections of A which contain B will be denoted $\mathrm{Cl}_r(B)$ and called the *r-closure* of B.

Example 1 The 'collection' of all adherents of a religion which bans mixed marriages may be said informally to be closed under the relation 'is married to'.

Exercise 1. Check that the r-closed subcollections of a collection A are the closed collections of a *topology* on A, in other words that:

$$\varnothing \text{ and } A \text{ are } r\text{-closed};$$

$$B, C\ r\text{-closed} \Rightarrow B \cup C\ r\text{-closed};$$

$$B_i\ r\text{-closed for all } i \in I \Rightarrow \bigcap_{i \in I} B_i\ r\text{-closed}.$$

Remark 1 In accordance with the convention described in remark 4 of §1.5 we shall sometimes write $\mathrm{Cl}_r(x)$ to denote $\mathrm{Cl}_r(\{x\})$.

Remark 2 $\mathrm{Cl}_r(B)$ is the smallest r-closed subcollection of A which contains B: it is said to be *r-generated* by B. Note that B is r-closed iff $\mathrm{Cl}_r(B) = B$.

Proposition 2.3.1

Suppose that r is a relation on A and $B \subseteq A$.
(a) $\mathrm{Cl}_r(B) = B \cup \mathrm{Cl}_r(r[B])$;
(b) $\mathrm{Cl}_r(r[B]) = r[\mathrm{Cl}_r(B)]$.

Proof (a) Clearly $B \subseteq \mathrm{Cl}_r(B)$. Moreover $r[B] \subseteq \mathrm{Cl}_r(B)$, and so $\mathrm{Cl}_r(r[B]) \subseteq \mathrm{Cl}_r(B)$. Hence $B \cup \mathrm{Cl}_r(r[B]) \subseteq \mathrm{Cl}_r(B)$. Now $r[\mathrm{Cl}_r(r[B])] \subseteq \mathrm{Cl}_r(r[B])$ and $r[B] \subseteq \mathrm{Cl}_r(r[B])$, so that

$$\begin{aligned} r[B \cup \mathrm{Cl}_r(r[B])] &= r[B] \cup r[\mathrm{Cl}_r(r[B])] \\ &\subseteq \mathrm{Cl}_r(r[B]) \\ &\subseteq B \cup \mathrm{Cl}_r(r[B]), \end{aligned}$$

and therefore $B \cup \mathrm{Cl}_r(r[B])$ is r-closed. But

$$B \subseteq B \cup \mathrm{Cl}_r(r[B]), \quad \text{and so} \quad \mathrm{Cl}_r(B) \subseteq B \cup \mathrm{Cl}_r(r[B]).$$

(b) $\mathrm{Cl}_r(B)$ is r-closed, hence so is $r[\mathrm{Cl}_r(B)]$. But $B \subseteq \mathrm{Cl}_r(B)$, so that $r[B] \subseteq r[\mathrm{Cl}_r(B)]$, and therefore $\mathrm{Cl}_r(r[B]) \subseteq r[\mathrm{Cl}_r(B)]$. Now $r[B] \subseteq \mathrm{Cl}_r(r[B])$ and $r[\mathrm{Cl}_r(r[B])] \subseteq \mathrm{Cl}_r(r[B])$, so that

$$\begin{aligned} r[\mathrm{Cl}_r(B)] &= r[B \cup \mathrm{Cl}_r(r[B])] \quad [(a)] \\ &= r[B] \cup r[\mathrm{Cl}_r(r[B])] \\ &\subseteq \mathrm{Cl}_r(r[B]). \end{aligned}$$

□

Definition

A relation r on A is said to be:
reflexive on A if $(\forall x \in A)(x\ r\ x)$;
irreflexive on A if $(\forall x \in A)(\text{not } x\ r\ x)$;
transitive if $(x\ r\ y \text{ and } y\ r\ z) \Rightarrow x\ r\ z$;
symmetric if $x\ r\ y \Rightarrow y\ r\ x$;
antisymmetric if $(x\ r\ y \text{ and } y\ r\ x) \Rightarrow x = y$.

Example 2 The largest relation on A is $A \times A$: it is reflexive, transitive, and symmetric, but not irreflexive unless $A = \varnothing$, nor antisymmetric unless A has at most one element.

Example 3 The diagonal id_A of A is reflexive, transitive, symmetric, and antisymmetric, but not irreflexive unless $A = \varnothing$.

Example 4 The smallest relation on A is $\varnothing$: it is transitive, symmetric, irreflexive, and antisymmetric, but not reflexive unless $A = \varnothing$.

Definition

If r is a relation on A then the intersection of the transitive relations on A containing r is called the *strict ancestral* of r and denoted r^{t}; the intersection of the reflexive transitive relations on A containing r is called the *ancestral* of r in A and denoted r^{T}.

The ancestral The ancestral is defined and its elementary properties are proved in Frege's *Begriffschrift* (1879). However, it was Dedekind's rediscovery of the notion in (1888) which led to its popularization.

Remark 3 It is easy to see that any intersection of [reflexive] transitive relations is also a [reflexive] transitive relation. It follows that the ancestral [respectively strict ancestral] of r is the smallest reflexive transitive relation [respectively transitive relation] on A containing r. The strict ancestral is consequently called the 'transitive closure' by some authors.

Example 5 Informally, if r is the relation 'is a parent of' on the collection of all humans, then r^t is the relation 'is an ancestor of' and r^T is the relation 'is the same person as or is an ancestor of'.

Proposition 2.3.2

(a) $r^T = r^t \cup \mathrm{id}_A$.
(b) $r \circ r^T = r^t = r^T \circ r$.

Proof (a) Note first that $r^t \subseteq r^T$ and that a relation on A is reflexive iff it contains id_A, so that $\mathrm{id}_A \subseteq r^T$. Thus $r \subseteq r^t \cup \mathrm{id}_A \subseteq r^T$. But $r^t \cup \mathrm{id}_A$ is transitive and reflexive, so that $r^t \cup \mathrm{id}_A = r^T$.

(b) Observe that $r \subseteq r \circ r^T$ since r^T is reflexive, and that $r \circ r^T$ is transitive, so that $r^t \subseteq r \circ r^T$. Moreover, if $x\ (r \circ r^T)\ z$ then there exists $y \in A$ such that $x\ r^T\ y$ and $y\ r\ z$: we then have either $x\ r^t\ y$ or $x = y$ [(a)], and in either case $x\ r^t\ z$. In other words $r \circ r^T \subseteq r^t$, whence $r \circ r^T = r^t$. A very similar argument shows that $r^t = r^T \circ r$. □

Proposition 2.3.3

(a) $r^T[B] = \mathrm{Cl}_r(B)$ *for all* $B \subseteq A$.
(b) $r^t[B] = r[\mathrm{Cl}_r(B)]$ *for all* $B \subseteq A$.

Proof (a) Suppose first that $x \in r^T[B]$ and $x\ r\ y$. Then there exists $a \in B$ such that $a\ r^T\ x$. Also $x\ r^T\ y$, and so $a\ r^T\ y$ by transitivity, i.e. $y \in r^T[B]$. This shows that $r^T[B]$ is r-closed, whence $\mathrm{Cl}_r(B) \subseteq r^T[B]$ since evidently $B \subseteq r^T[B]$ by the reflexivity of r^t. Now observe that the relation on A defined by the formula '$y \in \mathrm{Cl}_r(x)$' is reflexive and transitive and contains r: consequently r^T is contained in it. So if $y \in r^T[B]$ then there exists $x \in B$ such that $x\ r^T\ y$, hence $y \in \mathrm{Cl}_r(x) \subseteq \mathrm{Cl}_r(B)$. Therefore $r^T[B] \subseteq \mathrm{Cl}_r(B)$, and so $r^T[B] = \mathrm{Cl}_r(B)$.

(b)
$$\begin{aligned} r^t[B] &= (r \circ r^T)[B] \quad \text{[proposition 2.3.2(a)]} \\ &= r[r^T[B]] \\ &= r[\mathrm{Cl}_r(B)] \quad \text{[(a)]}. \end{aligned}$$
□

Corollary 2.3.4

(a) $x\ r^T\ y \Leftrightarrow \mathrm{Cl}_r(y) \subseteq \mathrm{Cl}_r(x)$.
(b) $x\ r^t\ y \Leftrightarrow \mathrm{Cl}_r(y) \subseteq r[\mathrm{Cl}_r(x)]$.

Proof (a)
$$\begin{aligned} x\ r^T\ y &\Leftrightarrow y \in r^T[x] \\ &\Leftrightarrow y \in \mathrm{Cl}_r(x) \quad \text{[proposition 2.3.3(a)]} \\ &\Leftrightarrow \mathrm{Cl}_r(y) \subseteq \mathrm{Cl}_r(x). \end{aligned}$$

(b)
$$x\, r^{t}\, y \Leftrightarrow y \in r^{t}[x]$$
$$\Leftrightarrow y \in r[\mathrm{Cl}_r(x)] \quad [\text{proposition 2.3.3(b)}]$$
$$\Leftrightarrow \mathrm{Cl}_r(y) \subseteq r[\mathrm{Cl}_r(x)] \quad [\text{proposition 2.3.1(b)}]. \qquad \square$$

Exercises 2. (a) Show that the smallest reflexive relation on A containing r is $r \cup \mathrm{id}_A$.
(b) Show that the smallest symmetric relation on A containing r is $r \cup r^{-1}$.
3. Show that a subcollection of A is r-closed iff it is r^{T}-closed.
4. If r is a relation on A, show that

$$r^{t} = \bigcap \{s \subseteq A \times A : r \cup (r \circ s) \subseteq s\}.$$

2.4 Equivalence relations

Definition

A relation on a collection A is called an *equivalence relation* if it is transitive, reflexive, and symmetric.

Remark 1 The conditions for s to be an equivalence relation on A may be expressed more compactly by the equation $s = (s \cup s^{-1})^{\mathrm{T}}$ (see exercise 2 below).

Example 1 The smallest equivalence relation on a collection A is the diagonal id_A; the largest is the whole product $A \times A$.

Example 2 The formula 'A and B are equinumerous' between collections A and B defines an equivalence relation on any day D.

Exercises 1. Show that a reflexive relation s on A is an equivalence relation iff

$$(x\ s\ y \text{ and } z\ s\ y \text{ and } z\ s\ t) \Rightarrow x\ s\ t.$$

2. If r is an arbitrary relation on A, show that the smallest equivalence relation on A containing r is $(r \cup r^{-1})^{\mathrm{T}}$.

Definition

The *equivalence classes* of an equivalence relation s on A are the collections $s[a] = \{x \in A : a\ s\ x\}$.

Warning The name 'equivalence class' is something of a misnomer as the collection in question need not be a class (in the sense which we shall introduce in Chapter 3), but it is hallowed by long usage.

Remark 2 The equivalence classes of s have the important property

$$(\forall a, b \in A)(a\ s\ b \Leftrightarrow s[a] = s[b]). \qquad (2.4.1)$$

When we are working locally, equivalence classes have much to commend them: you are no doubt familiar with the way in which consideration of equivalence classes provides us with a one-to-one correspondence between the equivalence relations on a collection and its partitions. Unfortunately, however, the equivalence classes of s are sometimes too big to be altogether satisfactory in more global work.

Exercise

3. If s is the equivalence relation on a limit day D defined by the formula 'a and b are equinumerous', show that the equivalence class $s[a]$ of an element a of D is not also an element of D unless a is empty; deduce that $\{s[a] : a \in D\}$ is not even contained in D.

It turns out that by using the cutting-down device introduced in §1.5 we can retain property (2.4.1) while reducing the size of the collections involved.

Definition

We define the *reduced equivalence classes* of an equivalence relation s on A to be the collections

$$s\langle a\rangle = \langle x \in A : a\ s\ x\rangle.$$

The collection $\{s\langle a\rangle\}_{a \in A}$ is denoted A/s and called the *quotient* of A by s; the corresponding function $(s\langle a\rangle)_{a \in A}$ is the graph of a surjective mapping $\mathrm{p}_s : A \to A/s$ which is called the *canonical quotient mapping*.

Remark 3 It is now an easy matter to check that if s is an equivalence relation then

$$x\ s\ y \Leftrightarrow s\langle x\rangle = s\langle y\rangle. \tag{2.4.2}$$

Remark 4 If s is an equivalence relation on a limit day D, then $D/s \subseteq D$.

If $f : A \to B$ is a mapping, then $\mathrm{Ker}[f]$ is an equivalence relation on A. The following proposition shows that every equivalence relation may be regarded as arising in this way.

Proposition 2.4.1

If s is an equivalence relation on A then $s = \mathrm{Ker}[\mathrm{p}_s]$.

Proof Immediate [(2.4.2)]. □

Proposition 2.4.2

If s is an equivalence relation on A then for every mapping $f : A \to C$ such that $s \subseteq \mathrm{Ker}[f]$ there is exactly one mapping $\bar{f} : A/s \to C$ such that the diagram

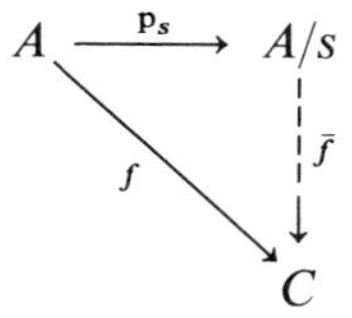

commutes.

Proof Immediate [propositions 2.2.4 and 2.4.1]. □

Exercises

4. Suppose that s is an equivalence relation on A.
   (a) $p_s^{-1}[p_s(x)] = s[x]$ for all $x \in A$.
   (b) Each element of A belongs to exactly one equivalence class.
5. If $f : A \to B$ is a mapping, show that there exist collections A' and B', an injective mapping $i : B' \to B$, a bijective mapping $f' : A' \to B'$, and a surjective mapping $p : A \to A'$ such that the diagram

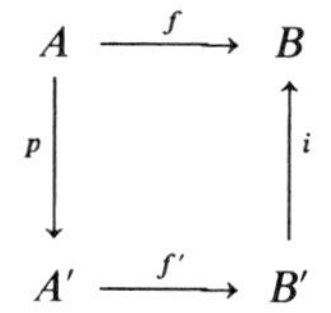

   commutes.

2.5 Structures

Definition

(A, r) is called a *structure* if r is a relation on A.

Remark 1 The structure (A, r) may be described as the result of *endowing* A with the relation r. Indeed there is a widely used convention that it may be denoted simply by A if the identity of the relation r is clear from the context. We shall make use of this convention occasionally.

Definition

Suppose that (A, r) and (B, s) are structures. A mapping $f : A \to B$ is said to be a *homomorphism* from (A, r) to (B, s) if $(\forall x, y \in A)(x\ r\ y \Rightarrow f(x)\ s\ f(y))$.

Remark 2 If $f : A \to B$ and $g : B \to C$ are homomorphisms, then it is clear that $g \circ f : A \to C$ is also a homomorphism. Moreover $\mathrm{id}_A : A \to A$ is trivially a homomorphism.

Definition

Suppose that (A, r) and (B, s) are structures. A bijective mapping $f : A \to B$ is called an *isomorphism* if both $f : A \to B$ and $f^{-1} : B \to A$ are homomorphisms.

We wish now to define 'substructures' and 'quotient structures': our guiding principle is that corollary 2.2.3 and proposition 2.3.2 should remain true when the collections are replaced by structures and the mappings by homomorphisms.

Definition

If (A, r) is a structure and $B \subseteq A$, then the relation $r \cap (B \times B)$ is denoted r_B and is said to be *induced* on B by r.

Proposition 2.5.1

If (A, r) is a structure and $B \subseteq A$, then r_B is the only relation on B such that the canonical inclusion mapping $\mathrm{id}_B : B \to A$ is a homomorphism and for every homomorphism $f : C \to A$ such that $\mathrm{im}[f] \subseteq B$ the unique mapping $\bar{f} : C \to B$ which makes the diagram

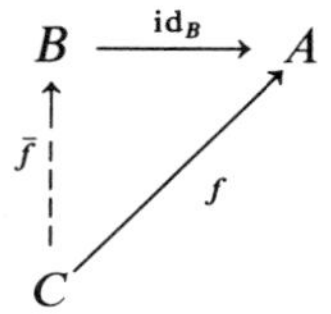

commute [corollary 2.2.3] is a homomorphism with respect to r_B.

Proof Straightforward. □

Definition

Suppose that (A, r) and (B, s) are structures. A mapping $f : A \to B$ is called an *embedding* if f is the graph of an isomorphism from (A, r) to $(f[A], s_{f[A]})$.

Remark 3 An isomorphism is the same thing as a surjective embedding.

Remark 4 Every embedding is an injective homomorphism. However, the converse need not hold: for example, if $r \subset s \subseteq A \times A$, then $\mathrm{id}_A : A \to A$ is a bijective homomorphism from (A, r) to (A, s) but it is clearly not an embedding.

Definition

If s is an equivalence relation on A and r is another relation on A, then by a slight abuse of notation we let r/s denote the relation on A/s given by $X\ (r/s)\ Y$ iff there exist $x, y \in A$ such that $X = s[x]$, $Y = s[y]$, and $x\ r\ y$.

Once again we obtain the desired mapping property:

Proposition 2.5.2

If (A, r) is a structure and s is an equivalence relation on A, then r/s is the only relation on A/s such that $\mathrm{p}_s : A \to A/s$ is a homomorphism and for *every homomorphism $f : A \to C$ such that $s \subseteq \mathrm{Ker}[f]$ the unique mapping $\bar{f} : A/s \to C$ which makes the diagram*

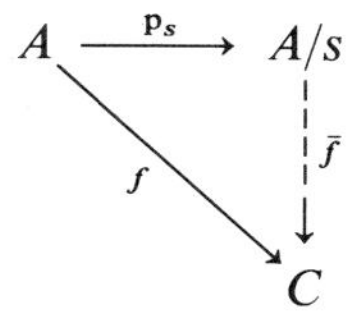

commute [proposition 2.3.2] is a homomorphism with respect to r/s.

Proof Straightforward. □

Now we turn to the definition of 'product' and 'coproduct' structures. Once again the definitions are forced on us by our desire to generalize the corresponding mapping properties for collections (propositions 2.4.4 and 2.4.5).

Definition

If $(A_i)_{i \in I}$, $(B_i)_{i \in I}$, and $(r_i)_{i \in I}$ are families of collections such that r_i is a relation between A_i and B_i for each $i \in I$, then (by a slight abuse of language) the relation from $\prod_{i \in I} A_i$ to $\prod_{i \in I} B_i$ defined by the formula $(\forall i \in I)(\mathrm{pr}_i(x)\ r_i\ \mathrm{pr}_i(y))$ is called the *product* of the relations r_i and denoted $\prod_{i \in I} r_i$.

Proposition 2.5.3

If $(A_i)_{i \in I}$ and $(r_i)_{i \in I}$ are families of collections such that r_i is a relation on A_i for each $i \in I$, then $\prod_{i \in I} r_i$ is the only relation on $\prod_{i \in I} A_i$ such that each $\mathrm{pr}_j : \prod_{i \in I} A_i \to A_j$ is a homomorphism and if (B, s) is a structure and $f_i : B \to A_i$ is a homomorphism for each $i \in I$ then the unique mapping $f : B \to \prod_{i \in I} A_i$ which makes the diagram

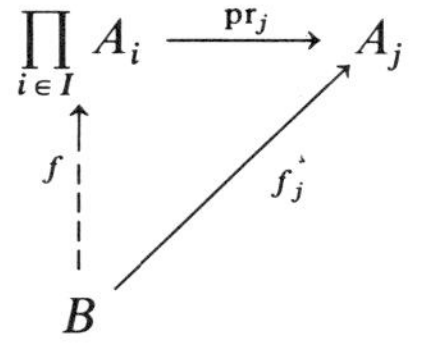

commute for all $j \in I$ [proposition 2.2.5] is a homomorphism with respect to $\prod_{i \in I} r_i$.

Proof Straightforward. □

Definition

Suppose that $(A_i)_{i \in I}$, $(B_i)_{i \in I}$, and $(r_i)_{i \in I}$ are families of collections suc that r_i is a relation between A_i and B_i for each $i \in I$. If we le $r_i' = \{((x, i), (y, i)) : x \; r_i \; y\}$, then (by a slight abuse of language) the relatio $\bigcup_{i \in I} r_i'$ between $\dot{\bigcup}_{i \in I} A_i$ and $\dot{\bigcup}_{i \in I} B_i$ is called the *direct sum* of th relations r_i and denoted $\dot{\bigcup}_{i \in I} r_i$.

Proposition 2.5.4

If $(A_i)_{i \in I}$ and $(r_i)_{i \in I}$ are families of collections such that r_i is a relatio on A_i for $i \in I$, then $\dot{\bigcup}_{i \in I} r_i$ is the only relation on $\dot{\bigcup}_{i \in I} A_i$ such tha each $\mathrm{p}_j : A_j \to \dot{\bigcup}_{i \in I} A_i$ is a homomorphism and if (B, s) is a structure an $f_i : A_i \to B$ is a homomorphism for each $i \in I$ then the unique mappin $f : \dot{\bigcup}_{i \in I} A_i \to B$ which makes the diagram

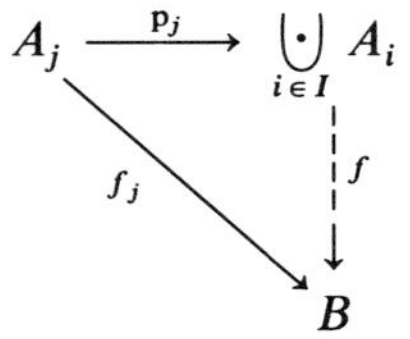

commute for all $j \in I$ [proposition 2.2.6] is a homomorphism with respec to $\dot{\bigcup}_{i \in I} r_i$.

Proof Straightforward. □

Exercises

1. Suppose that r is an equivalence relation on A.
   (a) If $B \subseteq A$, show that r_B is an equivalence relation on B and that B is r-close iff it is a union of equivalence classes.
   (b) If s is another equivalence relation on A, show that r/s is an equivalenc relation on A/s.
2. Suppose that $(A_i)_{i \in I}$ and $(s_i)_{i \in I}$ are families of collections such that s_i is a equivalence relation on A_i for each $i \in I$.
   (a) Show that $\Pi_{i \in I} s_i$ is an equivalence relation on $\Pi_{i \in I} A_i$.
   (b) Show that $\dot{\bigcup}_{i \in I} s_i$ is an equivalence relation on $\dot{\bigcup}_{i \in I} A_i$.

2.6 Order relations

Definition

A relation $\leq$ on a collection A is called a *partial ordering* and the pai $(A, \leq)$ is said to be *partially ordered* if the relation is transitive, anti symmetric, and reflexive. The relation is called a *total ordering* (or simpl an *ordering*) and the pair $(A, \leq)$ is said to be (*totally*) *ordered* if in addition:

$$(\forall x, y \in A)(x \leq y \text{ or } y \leq x).$$

Remark 1 Elements x and y of A are sometimes said to be *comparable* if either $x \leq y$ or $y \leq x$. With this terminology the partial ordering of A is total iff any two elements of A are comparable.

Remark 2 Suppose that $\leq$ is a partial [respectively total] ordering on A. If $B \subseteq A$ then $\leq_B$ is a partial [respectively total] ordering on B.

Remark 3 The symbol $\leq$ will be grossly overused in the rest of this book to denote almost every abstract partial ordering we have to refer to: in particular we shall denote $\leq_B$ by $\leq$ if confusion does not seem likely (and occasionally even if it does).

Remark 4 Suppose that $(A_i)_{i \in I}$ and $(\leq_i)_{i \in I}$ are families of collections such that $\leq_i$ is a partial ordering on A_i for $i \in I$. Then $\Pi_{i \in I} \leq_i$ is a partial ordering on $\Pi_{i \in I} A_i$. It is not a total ordering, however, unless at most one of the A_i has more than one element. Similarly $\dot{\bigcup}_{i \in I} \leq_i$ is a partial ordering of $\dot{\bigcup}_{i \in I} A_i$ which is a total ordering iff at most one of the A_i is non-empty and that one is totally ordered.

Definition

A relation $<$ on a collection A is called a *strict partial ordering* if it is transitive and irreflexive. It is called a *strict ordering* if in addition

$$(\forall x, y, z \in A)(x < y \text{ or } x = y \text{ or } y < x) \quad (\textit{trichotomy}).$$

Lemma 2.6.1

(*a*) *If* $\leq$ *is a [partial] ordering* $\leq$ *on* A *then the relation on* A *defined between* x *and* y *by the formula '*$x \leq y$ *and* $x \neq y$*' is a strict [partial] ordering on* A.

(*b*) *If* $<$ *is a strict [partial] ordering on* A *then the relation on* A *defined between* x *and* y *by the formula '*$x < y$ *or* $x = y$*' is a [partial] ordering on* A.

Proof Straightforward. □

Remark 5 Lemma 2.6.1 provides us with operations which associate a partial order with every strict partial order and conversely. It is easy to check that these two operations are inverses of one another. What this amounts to is that if we wish to define a partially ordered collection it is a matter of indifference to us whether it is the partial ordering or its associated strict partial ordering that we specify. (No relation on any but the empt collection can be both a partial ordering and a strict partial ordering.)

The inverse of the partial ordering $\leq$ is denoted $\geq$ and is also a partia ordering (total if $\leq$ is total). If we follow the convention referred to i remark 1 of §2.5 and denote the partially ordered collection $(A, \leq)$ simpl by the letter A, then the partially ordered collection $(A, \geq)$ is denoted A^{o} If $\Phi(A)$ is an assertion about the partially ordered collection A, the $\Phi(A^{\mathrm{op}})$ is called the *dual* assertion: if $\Phi(A) \Leftrightarrow \Phi(A^{\mathrm{op}})$ for all partiall ordered collections A, then we say that Φ is *self-dual*.

Exercise

1. If r is a relation on A and $s = (r \cap r^{-1})^{\mathrm{T}}$, show that s is an equivalence relatio on A and r^{T}/s is a partial ordering on A/s.

Suppose now that $(A, \leq)$ is a partially ordered collection. If $B \subseteq A$, the an element $a \in A$ is called a *lower bound* [respectively a *strict lower bound* for B if for every $x \in B$ we have $a \leq x$ [respectively $a < x$]. A lower boun for B which belongs to B is called a *least element* of B and is unique if i exists. An element $b \in B$ is *minimal in* B if there does not exist $x \in B$ suc that $x < b$. A least element of B must be minimal in B; the converse i true if $\leq$ is a total ordering, but not in general. [Strict] lower bounds an least and minimal elements with respect to the inverse partial orderin are called respectively [*strict*] *upper bounds* and *greatest* and *maxima* elements. B is said to be *bounded* in A if it has both an upper and a lowe bound in A.

Remark 6 The least [respectively greatest] element of a collection $B \subseteq A$ (if it exists is sometimes called the 'minimum' [respectively 'maximum'] of B an denoted $\min B$ [respectively $\max B$], despite the evident possibility o confusion between 'minimum' and 'minimal' [respectively 'maximum and 'maximal']. (For the record, minimum implies minimal, but no conversely: see exercise 2.) $\min A$ and $\max A$ are often denoted 0 and 1 although this notation is sometimes confusing and should be used wit caution. A widely used alternative, adopted here, is to denote them $\perp$ ('bottom') and $\top$ ('top') respectively.

The least upper bound [respectively greatest lower bound] of B (if it exists is called the *supremum* [respectively *infimum*] of B and denoted $\sup B$ [respectively $\inf B$].

The $\leq$-closure of B in A is denoted $\uparrow B$. Evidently

$$\uparrow B = \{x \in A : (\exists b \in B)(b \leq x)\}.$$

Certainly $B \subseteq \uparrow B \subseteq A$. If $B = \uparrow B$ (i.e. if B is $\leq$-closed in A), we say that B is *final* in A; if $\uparrow B = A$ (i.e. if B $\leq$-generates A), we say that B is *coinitial* in A. Dually, the $\geq$-closure of B is denoted $\downarrow B$, so that

$$\downarrow B = \{x \in A : (\exists b \in B)(b \geq x)\}.$$

B is *initial* in A if $B = \downarrow B$; B is *cofinal* in A if $\downarrow B = A$. The collection of all initial [respectively final] subcollections of A is denoted $\mathfrak{C}^{\downarrow}(A)$ [respectively $\mathfrak{C}^{\uparrow}(A)$].

Minimal elements of $\uparrow a - \{a\}$ are called *successors* of a; maximal elements of $\downarrow a - \{a\}$ are called *predecessors* of a. Note that a is a predecessor of b iff b is a successor of a. Successors of $\perp$ are called *atoms*; predecessors of $\top$ are called *coatoms*.

Definition

A non-empty partially ordered collection $(A, \leq)$ is said to be *directed* if $\{x, y\}$ has an upper bound in A for all $x, y \in A$. The dual notion is called a *filtered* collection.

Example 1 If $\mathscr{A}$ is a collection, then $\{a, b \in \mathscr{A} : a \subseteq b\}$ is a partial ordering on $\mathscr{A}$ which is (somewhat loosely) also denoted $\subseteq$. When we wish to regard $\mathscr{A}$ as a partially ordered collection in this way we refer to '$\mathscr{A}$ partially ordered by inclusion'; if $\mathscr{A}$ is totally ordered by inclusion then we call $\mathscr{A}$ a *chain*.

Example 2 A particular case of this is $(\mathfrak{P}(A), \subseteq)$: this partially ordered collection has a least element $\varnothing$ and a greatest element A; the supremum of a collection $\mathscr{B} \subseteq \mathfrak{P}(A)$ is $\bigcup \mathscr{B}$ and the infimum is $\bigcap \mathscr{B}$ if $\mathscr{B} \neq \varnothing$. $\mathfrak{P}(A)$ is a chain iff A has at most one element. The atoms of $\mathfrak{P}(A)$ are the singletons $\{a\}$ with $a \in A$.

Example 3 Another particular case is the collection $\mathfrak{P}(A, B)$ of *partial functions* from A to B, i.e. functions f such that $\mathrm{dom}[f] \subseteq A$ and $\mathrm{im}[f] \subseteq B$: the least element of this collection is the empty function; its maximal elements are the functions f such that $\mathrm{dom}[f] = A$. Every directed collection $\mathscr{B} \subseteq \mathfrak{P}(A, B)$ has a supremum in $\mathfrak{P}(A, B)$; in particular, every chain in $\mathfrak{P}(A, B)$ has a supremum. Every non-empty subcollection of $\mathfrak{P}(A, B)$ has an infimum.

Example 4 Any history $\mathscr{D}$ is a chain, i.e. it is totally ordered by inclusion. The associated strict ordering is that of membership [proposition 1.4.8(d)]. Note that every non-empty subclass of $\mathscr{D}$ has a least (i.e. earliest) element; we shall study ordered collections with this property in some detail in Chapter 6.

Exercises

2. (a) If a partially ordered collection $(A, \leq)$ has a least element, show that it is the unique minimal element of A.
   (b) Is it true, conversely, that if A has a unique minimal element then it is necessarily the least element of A?

3. (a) Show that every non-empty totally ordered collection is directed.
   (b) Show that every cofinal subcollection of a directed collection is directed.
   (c) Show that a maximal element of a directed collection is the greatest element.
4. Let $\mathrm{Po}(A)$ denote the collection of all partial orderings on A and let $\mathrm{Po}(A)$ be partially ordered by inclusion.
   (a) Prove that id_A is the least element of $\mathrm{Po}(A)$: in this context (i.e. when it is being thought of as a partial order) id_A is known as the *total unorder* relation on A.
   (b) Show that the maximal elements of $\mathrm{Po}(A)$ are precisely the total orderings on A.
5. If $(A, \leq)$ is a partially ordered collection and $B \subseteq A$, show that these two assertions are equivalent:
   (i) if $a \leq x \leq b$ and $a, b \in B$, then $x \in B$;
   (ii) $B = C \cap D$ where C is an initial subcollection and D is a final subcollection of A.

   We say that B is *convex* in A if it satisfies these conditions.
6. Let $\leq$ denote the ancestral of the relation on D defined between x and y by the formula '$x \in y$ and y is a collection'.
   (a) Show that $\leq$ is a partial ordering on D.
   (b) Show that $x \leq y \Rightarrow \mathrm{D}(x) \subseteq \mathrm{D}(y)$, but the converse implication may fail.
   (c) Show that every non-empty subcollection of D has a minimal element.
   (d) Show that these three assertions are equivalent:
      (i) A is initial in $(\mathrm{D}(A), \leq)$;
      (ii) $a \in A \Rightarrow a \subseteq A$;
      (iii) $\bigcup A \subseteq A$.

      We say that A is *transitive* in these circumstances.
   (e) Show that A is transitive iff $\mathfrak{P}(A) \cup \mathrm{D}_0$ is transitive.
   (f) The $\leq$-closure in $\mathrm{D}(A)$ of A is called the *transitive closure* of A and denoted $\mathrm{tc}(A)$. Show that $\mathrm{tc}(A)$ is the smallest transitive collection containing A. ($\mathrm{tc}(A) \cap \mathrm{D}_0$ is called the *footprint* of A; if it is empty, A is said to be *pure*.)
7. Retaining the notation of the previous exercise, let us call any pure transitive collection on which $\leq$ is a total ordering a *pseudo-ordinal*.
   (a) Show that a transitive collection is a pseudo-ordinal iff every element of it is a pseudo-ordinal.
   (b) If α and β are pseudo-ordinals in D, show that the following are equivalent:
      (i) $\alpha \in \beta$;
      (ii) $\alpha < \beta$;
      (iii) $\alpha \subset \beta$;
      (iv) $\mathrm{D}(\alpha) \in \mathrm{D}(\beta)$.
   (c) Deduce that every non-empty collection of pseudo-ordinals has a least element with respect to inclusion.
   (d) Show that $\varnothing$ is a pseudo-ordinal and that if α is a pseudo-ordinal then $\alpha^+ = \alpha \cup \{\alpha\}$ is a pseudo-ordinal (called the *successor* of α).
   (e) If D is a day, show that $\{\alpha \in D\colon \alpha \text{ is a pseudo-ordinal}\}$ is the unique pseudo-ordinal born on day D. It is called the *pseudo-rank* of D and denoted $\rho'(D)$.

(f) Show that $D(\rho'(D)) = D$ for every day D and $\rho'(D(\alpha)) = \alpha$ for every pseudo-ordinal α.
(g) α is called a *limit* pseudo-ordinal if $D(\alpha)$ is a limit day. Show that this is the case iff $\alpha \neq \varnothing$ and α is not the successor of any other pseudo-ordinal.

8. A pseudo-ordinal is said to be a *pseudo-natural* if neither it nor any of its elements is a limit pseudo-ordinal. Prove the following results.
(a) $\varnothing$ is a pseudo-natural.
(b) If n is a pseudo-natural then so is n^+.
(c) If $m^+ = n^+$ then $m = n$.
(d) $n^+ \neq \varnothing$ for any pseudo-natural n.
(e) If Φ is a formula such that $\Phi(\varnothing)$ and for every pseudo-natural n we have $\Phi(n) \Rightarrow \Phi(n^+)$, then we have $\Phi(n)$ for every pseudo-natural n.

9. Show how to define a term $m + n$ depending only on m and n such that:

if m, n are pseudo-naturals, then $m + n$ is a pseudo-natural;

$$m + n^+ = (m + n)^+;$$

$$m + \varnothing = m.$$

10. Show how to define a term mn depending only on m and n such that:

if m, n are pseudo-naturals, then mn is a pseudo-natural;

$$mn^+ = mn + m;$$

$$m\varnothing = \varnothing.$$

Definition

If $(A, \leq)$ and $(B, \leq)$ are partially ordered collections then a mapping $f: A \to B$ is said to be *increasing* (or *order-preserving*) if it is a homomorphism from $(A, \leq)$ to $(B, \leq)$, i.e. if $x \leq y \Rightarrow f(x) \leq f(y)$ for all $x, y \in A$, and *strictly increasing* (or *strictly order-preserving*) if it is a homomorphism from $(A, <)$ to $(B, <)$, i.e. if $x < y \Rightarrow f(x) < f(y)$ for all $x, y \in A$.

[Strictly] increasing mappings from A to B^{op} are said to be [*strictly*] *decreasing* (or [*strictly*] *order-reversing*) mappings from A to B. Mappings which are either [strictly] increasing or [strictly] decreasing are sometimes called [*strictly*] *monotonic*.

Remark 7 Notice that a mapping is an isomorphism between the partially ordered collections $(A, \leq)$ and $(B, \leq)$ iff it is an isomorphism between the associated strictly partially ordered collections $(A, <)$ and $(B, <)$.

Remark 8 Every embedding between partially ordered collections is increasing and injective, and every increasing injective mapping is strictly increasing; moreover, both converses are true if the domain is totally ordered, but neither converse is true in general (see exercise 11 below).

Exercise

11. Let $(A, \leq)$ and $(B, \leq)$ be partially ordered collections.
    (a) If $(A, \leq)$ is totally unordered (see exercise 4(a)), show that every mapping $A \to B$ is strictly increasing. Deduce examples of a strictly increasing bijective mapping which is not an isomorphism and a surjective strictly increasing mapping which is not bijective.
    (b) If $(A, \leq)$ is totally ordered, show that every strictly increasing mapping $A \to B$ is an embedding.

2.7 Complete and boundedly complete ordered collections

Lemma 2.7.1

Suppose that $(A, \leq)$ is a partially ordered collection.

(*a*) *These two assertions are equivalent*:
- (*i*) *every subcollection of A has a supremum in A*;
- (*ii*) *every subcollection of A has an infimum in A.*

(*b*) *These two assertions are equivalent*:
- (*i*) *every non-empty subcollection of A which has an upper bound in A has a supremum*;
- (*ii*) *every non-empty subcollection of A which has a lower bound in A has an infimum.*

Proof We shall prove (a) and leave (b) as an exercise. Suppose that $B \subseteq A$ and let $C = \{x \in A : x \text{ is a lower bound for } B\}$. By hypothesis $a = \sup C$ exists. We shall show that $a = \inf B$. To do this note first that if $x \in C$ and $y \in B$ then $x \leq y$: consequently y is an upper bound for C and so $a \leq y$. In other words, a is a lower bound for B. And if x is another lower bound for B then $x \in C$ and so $x \leq a$. Thus a is the greatest lower bound for B, i.e. $a = \inf B$. This proves that (i) $\Rightarrow$ (ii); to prove that (ii) $\Rightarrow$ (i) apply the implication (i) $\Rightarrow$ (ii) to A^{op}. □

Definition

A partially ordered collection $(A, \leq)$ is said to be *complete* if it satisfies the equivalent conditions of lemma 2.7.1(a), and *boundedly complete* if it satisfies the conditions of lemma 2.7.1(b).

Remark 1 A partially ordered collection is complete iff it is both boundedly complete and bounded.

Example 1 $\mathfrak{P}(A)$ is complete with respect to inclusion.

Example 2 The collection $\mathfrak{P}(A, B)$ of partial functions from A to B is boundedly complete, but is not complete with respect to inclusion if B has more than one element and A is non-empty, since then the union of a non-directed collection of functions need not be contained in a function. However, we can make $\mathfrak{P}(A, B)$ complete by adjoining a top element. This example is important in the theory of computing.

Example 3 If $(A, \leq)$ is a partially ordered collection then the collection $\mathfrak{C}^{\downarrow}(A)$ of initial subcollections of A is complete with respect to inclusion: the supremum and infimum of a subcollection of $\mathfrak{C}^{\downarrow}(A)$ are its union and intersection respectively. Also the mapping $j: A \to \mathfrak{C}^{\downarrow}(A)$ given by $a \mapsto \downarrow a$ is an order-embedding. The intersection of all the complete subcollections of $\mathfrak{C}^{\downarrow}(A)$ containing $j[A]$ is denoted $\bar{A}$; the ordered pair $(\bar{A}, j)$ is called the *completion* of A.

Exercises

1. Prove lemma 2.7.1(b).
2. (a) Show that the intersection of a family of equivalence relations is one too.
   (b) Deduce that the collection $\mathrm{Equ}(A)$ of equivalence relations on a collection A is complete with respect to inclusion.
   (c) Show that the union of even just two equivalence relations need not be one.
   (d) Deduce that the supremum of a subcollection of $\mathrm{Equ}(A)$ is not in general its union.
3. (a) Show that the intersection of a non-empty collection of partial orderings is also a partial ordering.
   (b) Deduce that $\mathrm{Po}(A)$ is boundedly complete with respect to inclusion.
   (c) Show that $\mathrm{Po}(A)$ is complete iff A has no more than one element.
4. If A is boundedly complete, show that $\bar{A}$ is isomorphic to A with bottom and top elements adjoined if necessary.

Definition

If $\mathscr{F}$ is a collection of functions on A then we let $\mathrm{fix}(\mathscr{F}) = \{a \in A : f(a) = a$ for all $f \in \mathscr{F}\}$. The elements of $\mathrm{fix}(\mathscr{F})$ are called *fixed points* of $\mathscr{F}$.

Theorem 2.7.2 (Knaster 1927)

Suppose that $(A, \leq)$ is complete. If $f: A \to A$ is increasing then f has a least fixed point

$$\min \mathrm{fix}(f) = \min\{x \in A : f(x) \leq x\}.$$

Proof Let $B = \{x \in A : f(x) \leq x\}$ and let $a = \inf B$. Then for all $x \in B$ we have $a \leq x$ and hence $f(a) \leq f(x) \leq x$. Thus $f(a)$ is a lower bound for B and so $f(a) \leq a$; hence $f(f(a)) \leq f(a)$, i.e. $f(a) \in B$, and so $f(a) \geq a$. Therefore $f(a) = a$. Finally, if a' is another fixed point of f, then $a' \in B$ and so $a \leq a'$. □

Exercises

5. If $(A, \leq)$ is complete and $\mathscr{F}$ is a collection of increasing mappings $A \to A$ such that $f \circ g = g \circ f$ for all $f, g \in \mathscr{F}$, show that $\mathrm{fix}(\mathscr{F})$ has a least element

$$\min \mathrm{fix}(\mathscr{F}) = \min\{x \in A : f(x) \leq x \text{ for all } f \in \mathscr{F}\}.$$

6. Suppose that $(A, \leq)$ is complete and $f: A \to A$ is increasing.
   (a) Show that if $x \in \mathrm{Cl}_f(\bot)$ then $f(x) \geq x$.
   (b) Show that the function $A \to A$ given by $x \mapsto \sup \mathrm{Cl}_f(x)$ is increasing.

Definition

Suppose that $(A, \leq)$ and $(B, \leq)$ are partially ordered collections. A mapping $f: A \to B$ is said to be *normal* if for every collection C which has a supremum in A the collection f$[C]$ has a supremum in B and f$(\sup C) = \sup$ f$[C]$; it is said to be *strictly normal* if in addition it is injective.

Remark 2 Every [strictly] normal mapping is in particular [strictly] increasing.

Exercise

7. Suppose that $(A, \leq)$ is complete.
   (a) If $f: A \to A$ is normal, show that
   $$\min \operatorname{fix}(f) = \sup \operatorname{Cl}_f(\bot).$$
   (b) If $f, g: A \to A$ are normal, show that
   $$\min \operatorname{fix}(f \circ g) = f(\min \operatorname{fix}(g \circ f)).$$
   (c) If $f, g: A \to A$ are normal and $f \circ g = g \circ f$, show that
   $$\min \operatorname{fix}(f, g) = \sup \operatorname{Cl}_{f,g}(\bot) = \min \operatorname{fix}(f \circ g)$$
   (where $\operatorname{Cl}_{f,g}(\bot)$ is the smallest subcollection of A containing $\bot$ which is both f-closed and g-closed).
   (d) If in addition $f(\bot) = g(\bot)$, show that $\min \operatorname{fix}(f) = \min \operatorname{fix}(g)$.

2.8 Successor algebras

In the next chapter we shall develop the properties of the natural numbers. In the meantime let us introduce the concept on which our treatment of the natural numbers will be based.

Definition

A structure (A, f) is called a *successor algebra* if f is a function from A to A.

Example Any limit day D becomes a successor algebra when it is endowed with the function $x \mapsto x \cup \{x\}$.

Since successor algebras are structures, the definitions of §2.5 may be applied to them. In particular, if (A, f) and (B, g) are successor algebras, then a mapping $u: A \to B$ is a homomorphism iff

$$x \; f \; y \Rightarrow u(x) \; g \; u(y),$$

i.e. iff

$$u(f(x)) = g(u(x)) \tag{2.8.1}$$

for all x, $y \in A$.

Remark Between successor algebras every injective homomorphism is an embedding and every bijective one is an isomorphism (see exercise 1(b) below). This is in contrast to what is true for arbitrary structures (see remark 4 of §2.5), but is typical of algebraic situations (e.g. groups, rings, vector spaces, etc.).

Exercises

1. Suppose that (A, f) and (B, g) are successor algebras.
   (a) Show that a mapping $u: A \to B$ is a homomorphism iff its graph u is $(f \times g)$-closed in $A \times B$.
   (b) Show that a mapping $u: A \to B$ is an embedding iff it is an injective homomorphism.
2. If (A, f) is a successor algebra and $B \subseteq A$, show that (B, f_B) is a successor algebra iff B is f-closed in A. (In these circumstances we say that B is a *subalgebra* of A.)
3. If $((A_i, f_i))_{i \in I}$ is a family of successor algebras, show that $(\prod_{i \in I} A_i, \prod_{i \in I} f_i)$ and $(\dot\bigcup_{i \in I} A_i, \dot\bigcup_{i \in I} f_i)$ are successor algebras.
4. If (A, f) is a successor algebra and s is an equivalence relation on A, find a necessary and sufficient condition on s for $(A/s, f/s)$ to be a successor algebra.
5. If D is a limit day, show that the function $x \mapsto x \cup \{x\}$ referred to in the example above is the graph of an injective mapping $D \to D$ and that $\varnothing$ does not belong to its image.

Lemma 2.8.1

Suppose that (A, f) and (B, g) are successor algebras. If u, $v: A \to B$ are homomorphisms, then $\{x \in A : u(x) = v(x)\}$ is f-closed.

Proof Let $C = \{x \in A : u(x) = v(x)\}$. Then

$$\begin{aligned} x \in C &\Rightarrow u(x) = v(x) \\ &\Rightarrow u(f(x)) = g(u(x)) = g(v(x)) = v(f(x)) \quad [(2.8.1)] \\ &\Rightarrow f(x) \in C \end{aligned}$$

and so C is f-closed. □

3 Basic set theory

3.1 Sets and classes

None of the axioms we have stated so far prevents the possibility that all collections are finite. So if we want to be able to construct infinite collections in our theory—which we do—we must make the further assumption that such collections exist. There are two ways of doing this: we can assert either that there are infinitely many individuals (that, in the terminology suggested by Fig. 1, the universe is infinitely wide) or that there are infinitely many days (that the universe is infinitely long). We shall take the latter path here: we can achieve what we want by asserting the existence of at least one limit day. (The definition of 'infinite' will not be given until Chapter 4; in the meantime we rely on naive notions for our discussion.) With this addition we obtain a system called **ZA** which is adequate for the presentation of much of modern mathematics.

However, we shall for convenience make a further (permissive) change: we shall postulate the existence of a univeral class. We cannot assert the existence of a collection to which every collection belongs, since it would belong to itself, contradicting proposition 1.5.1. There is nevertheless an undoubted appeal in the notion of having at our disposal a particular collection which, while not containing *all* collections, does contain enough to cope with the needs of everyday mathematics. A convenient way to do this is to add to our language a constant **D** and to postulate that **D** is a limit day but not the earliest one.

Explicitly, we assume from now on that our language contains a proper name **D** to denote an object called the *universal class*: its members are said to be *small*; small collections are called *sets*; subcollections of **D** are called *classes*; classes which are not sets are called *proper classes.*

Power-set axiom

D *is a limit day.*

The power-set axiom is equivalent to the assertion that the power of any set is a set (hence the name).

Remark 1 Every set is a class and every class is a collection, but neither converse holds: **D** is a class which is not a set and $\{\mathbf{D}\}$ is a collection which is not a class.

Remark 2 The use of 'small' as the adjective associated with the noun 'set' is not ideal: $\{\mathbf{D}\}$ only has one element but is not small. 'Antediluvian' would accord better with our other terminology, but this use of 'small' is now well established (particularly in works on category theory) and it is therefore probably best to conform.

Remark 3 Since **D** is in particular a day and hence a collection, we can now drop the temporary axiom asserting the existence of at least one collection.

Proposition 3.1.1

(*a*) $a \subseteq \mathbf{D} \Rightarrow \mathrm{cut}(a) \in \mathbf{D}$.
(*b*) $a \in \mathbf{D} \Rightarrow \mathfrak{P}(a) \in \mathbf{D}$.
(*c*) $a \in \mathbf{D} \Rightarrow \bigcup a \in \mathbf{D}$.
(*d*) $x, y \in \mathbf{D} \Rightarrow \{x, y\} \in \mathbf{D}$.
(*e*) $x, y \in \mathbf{D} \Rightarrow (x, y) \in \mathbf{D}$.
(*f*) $a, b \in \mathbf{D} \Rightarrow a \times b \in \mathbf{D}$.

Proof Immediate [power-set axiom and propositions 1.5.3, 1.5.8, and 2.1.3]. □

Axiom of infinity

There is a limit day earlier than **D**.

The earliest limit day is denoted D_ω: the axiom of infinity asserts that D_ω is a set. When we have defined what it is for a collection to be 'infinite' (§4.3) it will be apparent that the axiom of infinity is equivalent to the assertion that there exists an infinite set (hence the name).

Infinite collections Our intuitions about infinite collections are undoubtedly more nebulous and more capable of arousing controversy than those about finite ones. Indeed many of the properties of infinite collections seem paradoxical at first sight. When Cantor wrote his works on set theory, an abhorrence of actually infinite collections had been commonplace since Aristotle: they were, for example, scrupulously avoided by Euclid (*c*.300 BC). There were certainly a few heretics—Galileo and Bolzano, for example—but the opinions of Gauß were no doubt particularly influential in nineteenth-century Germany (letter to Schumacher, 1831): 'I protest ... against using infinite magnitudes as something consummated; such a use is never admissible in mathematics.' It was just at the time when infinitesimals had been successfully expunged from analysis by Weierstraß (Cantor's teacher)—not to be rehabilitated until the 1950s—that Cantor himself tamed the paradoxes of the actual infinite in his work on cardinal and ordinal arithmetic. However, he had to brook considerable opposition and devoted lengthy passages in his published work to the defence of his views (1932, p. 374): 'The fear of infinity is a form of myopia which destroys the possiblity of seeing the actual infinite.'

Even after the fruitfulness of Cantor's exploitation of the infinite had begun to win it converts, it took some time for it to be realized that the existence of infinite collections is not logically inevitable. Frege's attempt to prove their existence failed because it depended crucially on the fatally flawed type-collapsing which we have already referred to. Even as late as 1903 Russell in *The principles of mathematics* made the remarkable statement: 'That there are infinite classes is so evident that it will scarcely be denied. Since, however, it is capable of formal proof, it may be as well to prove it.' (Russell 1903, p. 357.) (The proof he offers is, of course, fallacious.)

Exercise Show that a is a proper class iff $\mathrm{D}(a) = \mathbf{D}$.

The theory whose axioms are SUBCOLLECTIONS, creation, power-set, and infinity is called basic set theory *or* **GA**. *Except where we state the contrary explicitly, we shall work in* **GA** *throughout the remainder of this book.*

3.2 Individuals

We are developing a theory of collections: but collections of what? Presumably the intention is that we should start with a domain of things—numbers, points, torsion-free divisible groups, Buddhists—which are to be the object of study and which we want to be able to form into collections. We call these things, whatever they are, 'individuals'. From the point of view we adopt in this book—that of the collection-theorist—what they are is in any case irrelevant since nothing can be said in the language of collection theory about their internal structure. Let us now express this more formally.

Definition

If a is a collection and Φ is a formula which does not depend on a, then the *relativization* of Φ to a is the formula $\Phi^{(a)}$ obtained from Φ by replacing every quantifier $(\forall x)$ or $(\exists x)$ by $(\forall x \in a)$ or $(\exists x \in a)$ respectively.

Example 1 $((\exists! x)\Phi(x))^{(a)}$ is the formula $(\exists x \in a)(\forall y \in a)(\Phi^{(a)}(y) \Leftrightarrow x = y)$. It is read 'There is exactly one $x \in a$ such that $\Phi^{(a)}(x)$'.

Suppose that $\mathscr{T}$ is a first-order mathematical theory. Consider the theory obtained from basic set theory by adding as axioms the relativizations of all the axioms of $\mathscr{T}$ to the collection D_0 of individuals. This theory is called 'Basic set theory applied to $\mathscr{T}$' and denoted $\mathrm{GA}[\mathscr{T}]$.

Example 2 The *language of arithmetic* has one binary predicate succ(x, y) and two ternary predicates add(x, y, z) and mult(x, y, z). We define the following terms:

$$0 = \iota! y(\forall x) \text{ not succ}(x, y)$$

$$x^+ = \iota! y \text{ succ}(x, y)$$

$$x + y = \iota! z \text{ add}(x, y, z)$$

$$x \cdot y = \iota! z \text{ mult}(x, y, z).$$

PA (*Peano arithmetic*) is the theory in this language with the following axioms:

$$0 \text{ exists}$$

$$(\forall x)(x^+ \text{ exists})$$

$$(\forall x)(\forall y)(x + y \text{ exists})$$

$$(\forall x)(\forall y)(x \cdot y \text{ exists})$$

$$(\forall x)(\forall y)(x^+ = y^+ \Rightarrow x = y)$$

$$(\forall x)(x + 0 = x)$$

$$(\forall x)(\forall y)(x + y^+ = (x + y)^+)$$

$$(\forall x)(x \cdot 0 = 0)$$

$$(\forall x)(\forall y)(x \cdot y^+ = x \cdot y + x).$$

If Φ is any formula in the language of arithmetic, then

$$(\Phi(0) \text{ and } (\forall x)(\Phi(x) \Rightarrow \Phi(x^+))) \Rightarrow (\forall x)\Phi(x).$$

All the familiar arithmetical facts about natural numbers are provable in **PA**: the details are in Chapter I of Landau's *Grundlagen der Analysis* (1930), for instance. Whether *all* arithmetical facts about natural numbers are provable in **PA** is a question to which we shall return later. We will show in the next chapter that GA[**PA**] is in fact no stronger than **GA**.

Example 3 An extreme case is the theory **null** which has no axioms at all. GA[**null**] evidently coincides with **GA**.

Example 4 Another extreme case is the theory **empty** in which $(\forall x)(x \neq x)$ is provable. The only model of this theory is empty (hence the name). GA[**empty**] is equivalent to the theory **G** we obtain from **GA** by adding the *axiom of purity*, i.e. the assertion that there are no individuals.

Inclusive logic According to the traditional presentation of first-order logic **empty** is inconsistent, but this is widely regarded as a 'defect in logical purity' (Russell 1919, p. 203n.). The study of systems of logic which do not make this assumption ('inclusive' logics) was initiated by Jaskowski in 1934 and has been pursued by Hailperin (1953), Quine (1954), and others.

(Easily accessible accounts of such systems are given in Hodges (1977) and Johnstone (1987).)

Purity It is not difficult to show that the axiom of purity is consistent with **GA**. Like the other restrictive hypothesis we have considered, the birthday principle, the axiom of purity is rarely used outside set theory proper. But it is appropriate only to the study of a vacuum and is therefore, unlike the birthday principle, abhorrent to any natural conception of mathematics.

3.3 Summary

In this section we shall review the properties of **GA** and compare it with other axiomatizations which you may encounter. It will be helpful to distinguish three types of theory in common use, which we shall (with at least some historical justification) call Zermelo, von Neumann, and Grothendieck theories. In Zermelo theories the creation process stops just *before* it reaches **D**, so that the only collections in the theory are sets; in von Neumann theories the creation process stops *at* **D**, so all collections are classes; in this book we have decided, for reasons of convenience, to use a Grothendieck theory which carries on *beyond* **D** to create collections which are not classes. To make the axiom system **GA** we have given here into a von Neumann-type system **VA**, replace the axiom of creation by '$(\forall a)(a \subseteq \mathbf{D})$' and restrict the AXIOM OF SUBCOLLECTIONS so that the variables other than x on which $\Phi(x)$ depends range only over small collections. (If we do not restrict the AXIOM OF SUBCOLLECTIONS in this way, we obtain a slightly stronger system of a type to be found in Morse (1965) and Kelley (1955).) To obtain a Zermelo-type system **ZA**, on the other hand, delete all mention of the constant **D** (including the power-set and infinity axioms) and instead take as the axiom of infinity $(\exists \mathrm{D})(\mathrm{D}$ is a limit$)$. The three types of system are illustrated in Fig. 1.

If we are only interested in proving things about sets, then the choice between a Zermelo and a von Neumann theory (e.g. between **ZA** and **VA**) is entirely one of personal preference because exactly the same things are provable about sets in each. To be precise, if Φ is a first-order set-theoretic sentence in which the proper name **D** does not occur, then Φ is provable in **ZA** iff $\Phi^{(\mathbf{D})}$ is provable in **VA**. However, with Grothendieck theories things are quite different: there are sentences Φ such that $\Phi^{(\mathbf{D})}$ is provable in **GA** but Φ is *not* provable in **ZA** (Kreisel and Levy 1968, theorems 10 and 11).

Now if we take our account of the stage-by-stage creation of collections seriously then it is at first sight bizarre that a decision to carry on creating more collections should affect what it is possible to prove about collections which have already been created. The reason for this apparent

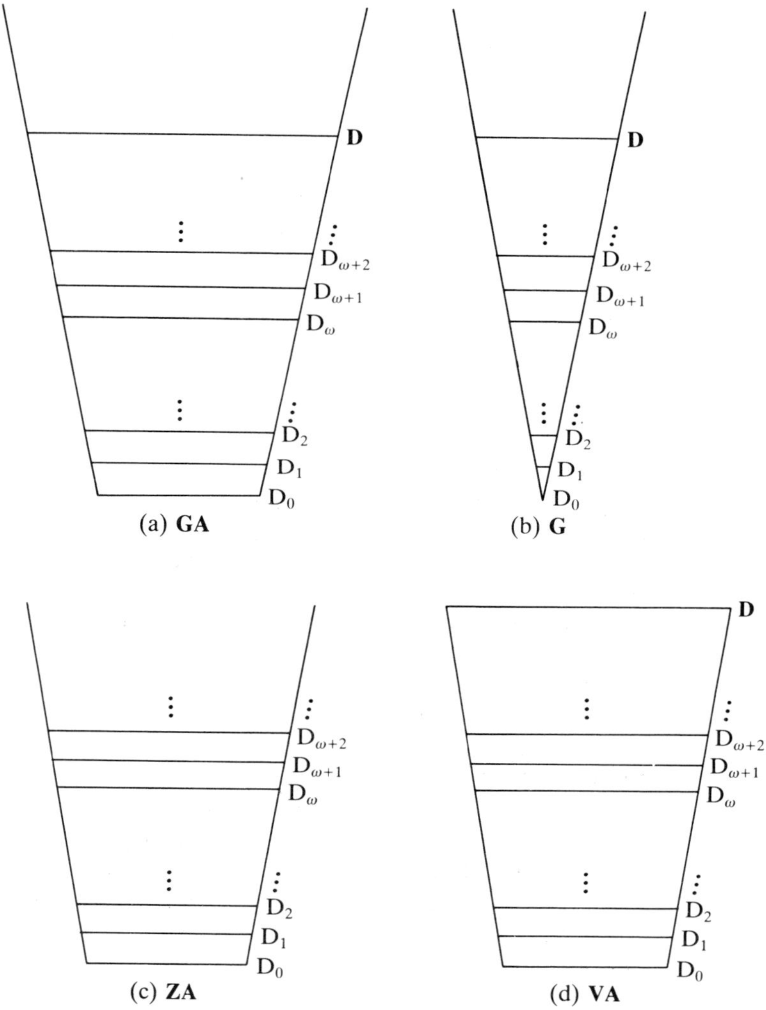

Figure 1

anomaly is the impredicativity in the AXIOM OF SUBCOLLECTIONS which we have already mentioned: if we let $a_y = \{x \in a : \Phi(x, y)\}$ then a_y exists for *all* collections y, even those which are created *after* a; if we strengthen the axiom of infinity we guarantee the existence of more collections y and hence perhaps of more subcollections a_y of a.

Zermelo, von Neumann, and Grothendieck theories Cantor's letters to Dedekind (1899*a*, *b*) contain explicit mention of a distinction between the consistent multiplicities called 'sets' and the inconsistent or 'absolutely infinite' ones. And Zermelo's (1908*a*) paper on the axiomatization of set theory begins: 'Set theory is concerned with a *domain* $\mathfrak{B}$ of individuals, which we shall call simply *objects* and among which are the *sets*. We say of an object a that it "exists" if it belongs to the domain $\mathfrak{B}$; likewise we say of a class $\mathfrak{K}$ of objects that "there exist objects of the class $\mathfrak{K}$" if $\mathfrak{B}$ contains at least one individual of this class.' Later he observes that 'the domain $\mathfrak{B}$ is not itself a set'. It is clear, therefore, that Zermelo did not envisage a situation in which classes which are not sets could not be mentioned at all. However, he did not make this precise. As a result many mathematicians regarded Zermelo's axiomatization as banning direct reference to proper classes and used the set-theoretic paradoxes as a justification for their fear of directly mentioning the absolute in much the same way as previous generations had used the paradoxes of the infinite as a justification for banning reference to the actual infinite.

It is only much more recently that it has been shown how talk about proper classes may be rigorously interpreted in Zermelo's theory. According to Quine (1963), for instance, any mention of a proper class is to be regarded as a mere *façon de parler* which can be eliminated by rewriting the sentence in which the mention occurs. But it was regarded as a new idea when in 1925 (still before Cantor's letters to Dedekind had been published) von Neumann produced a formalized theory which rehabilitated the universal class.

The notion that it may be useful to go not just as far as the universal class **D** but beyond it has grown more recently, principally in response to the needs of category theory, and in this context is usually connected with the name of Grothendieck (see MacLane 1971, p. 24; Hayden and Kennison 1968). However, it is scarcely a surprising idea—why stop at **D**?—and had doubtless appeared elsewhere before Grothendieck. (Of course, if we took seriously the religious metaphor whereby the universal class **D** is regarded as the absolute being then the idea of proceeding beyond **D** would no doubt seem heretical.)

Whichever version of basic set theory we choose, sets have the following properties

Property 1

If a, b are sets and $(\forall x)(x \in a \Leftrightarrow x \in b)$, then $a = b$.

Property 2

(a) $\varnothing$ is a set.
(b) If a is a set then $\{a\}$ is a set.
(c) If a, b are sets, then $\{a, b\}$ is a set.

Property 3

If a is a set and $\Phi(x)$ is a formula, then $\{x \in a : \Phi(x)\}$ is a set.

Property 4

If a is a set, then $\mathfrak{P}(a)$ is a set.

Property 5

If a is a set, then $\bigcup a$ is a set.

Property 6

(a) There is a set a such that $\varnothing \in a$ and $x \in a \Rightarrow (x) \in a$.
(b) There is a set a such that $\varnothing \in a$ and $x, y \in a \Rightarrow x \cup \{y\} \in a$.

Property 7

If a is a set which has sets as elements, then it has an element b such that $a \cap b$ has no sets as elements.

Zermelo's (1908*a*) axiom system consisted of properties 1–6(a) together with the axiom of choice (see Chapter 7). If we add property 7 and use the rather stronger version of the axiom of infinity contained in property 6(b), then we obtain another formulation of **ZA**. The theory we obtain from **ZA** by adding the axiom of purity is called just **Z**. As we have already remarked, the system **GA** which we have adopted in this book is slightly stronger than **ZA**. It has become common for mathematicians to add to **Z** [respectively **ZA**] the *AXIOM OF REPLACEMENT*, which asserts that if a is a set, $\tau(x)$ is a set for all $x \in a$, and the variables in $\tau(x)$ other than x (if any) range over sets, then $\{\tau(x) : x \in a\}$ is a set. The resulting theory is called **ZF** [respectively **ZFA**] (F for 'Fraenkel'). **ZF** has achieved a rather dominant position among mathematicians as the standard foundation for their subject.

Replacement Cantor's published works contain several assertions which are only true if the universe is rather large: for example, he claimed in 1883 that there is an aleph for every ordinal (cf. §6.6), although the assumptions which he made explicitly there do not imply this. It was only in a letter to Dedekind (1899*a*, not published until 1932) that he stated, as part of a response to the set-theoretic paradoxes which he had fairly recently discovered, a property strong enough to justify his claims: if two classes are the same size then either both are sets or neither is ('the doctrine of limitation of size'). This assumption is motivated by the view that the reason why some classes are not sets is that they are too big. It appeared again, stated informally, in Mirimanoff (1917), Lennes (1922), and Fraenkel (1922*a*), and was made precise by Skolem (1922); the name 'axiom of replacement' (*Ersetzungsaxiom*) under which it is nowadays known is due to Fraenkel.

The fact that I have omitted REPLACEMENT from the system of basic set theory which we shall use in the remainder of this book should not be taken to indicate authorial disapproval. Nor should it encourage you to think that it may be inconsistent: only a believer in a dice-playing God could think that. Rather is it evidence of a dislike of unnecessary assumptions, particularly ones which are not needed in any but the most esoteric parts of mathematics: only one result outside set theory—the assertion that every Borel game is determined—has so far been shown to require anything approaching the strength of REPLACEMENT for its proof (Friedman 1971; Martin 1975). Even von Neumann, who used REPLACEMENT frequently (and vitally) in his set theory, admitted that it 'goes a bit too far' (1925, p. 227).

Are the axioms true? For Euclid an axiom was a 'self-evident truth'. However, the part of mathematics we shall be studying here is much more abstract than those—arithmetic and affine geometry—with which Euclid dealt, and our intuitions about it are correspondingly more difficult to clarify. So it would be going a little far—particularly in view of the historical development of set theory—to describe the axioms in this book as self-evident. Nevertheless, I do wish to claim that I have an intuitive mental picture of the system we shall be studying and that all the axioms of **GA** are, when interpreted as statements about this system, assertions which I believe to be true. They are most emphatically not just 'a set of arbitrarily chosen statements which, together with the rule of *modus ponens*, suffice to derive all the statements which we wish to derive' (Rosser 1953, p. 55). (For a different view read MacLane (1986, pp. 440–9).)

Of course this is not to suggest that one cannot picture other systems of set theory which some mathematicians may regard as more appropriate or more interesting objects of study: set theory is a house with many mansions just as geometry is. Nor is it to suggest that my belief in the truth of the axioms may not be mistaken in the same way that anyone describing an out-of-focus picture can be mistaken. However, if the axioms were arbitrarily chosen then the presentation of their logical consequences, while it would involve the same problems of checking for formal correctness, would scarcely be a matter in which I could expect you to be interested. Jourdain had it right (1906, p. 270n.): 'It cannot, I think, be maintained as an historical fact that any advance in mathematics has been brought about by an arbitrary creation of the mind, of which the fruitfulness has only been discovered afterwards, but one cannot object, on logical grounds, to such a creation if non-contradictory.'

3.4 The axiom of indescribability*

One of the things the axioms of basic set theory are designed to ensure is that the universal class **D** is stable under the everyday operations of mathematics—the formation of power sets, cartesian product sets, infinite sets, etc. However, they are not sufficient to ensure that **D** is stable under *all* set-theoretic operations. For example, it is impossible to prove in basic set theory that $\bigcup\{D_\omega, \mathfrak{P}(D_\omega), \mathfrak{P}(\mathfrak{P}(D_\omega)), \ldots\}$ is a set, despite the fact that all the objects from which this class is constructed—D_ω, $\mathfrak{P}(D_\omega)$, etc.—are sets. The AXIOM OF REPLACEMENT overcomes this particular problem but seems somewhat arbitrary: why should **D** be stable under some operations but not others? The boldest stroke would be to assume that every collection which is definable purely in terms of sets is itself a set. Put more formally:

AXIOM OF INDESCRIBABILITY

If $\tau(x_1, \ldots, x_n)$ is a term depending on $x_1, \ldots, x_n$ in which the constant **D** *does not occur, then this is an axiom:*

$$(\forall x_1, \ldots, x_n \in \mathbf{D})(\tau(x_1, \ldots, x_n) \text{ exists} \Rightarrow \tau(x_1, \ldots, x_n) \in \mathbf{D}).$$

We shall assume the AXIOM OF INDESCRIBABILITY only for the rest of this section: to emphasize this we place a dagger † beside the statements of results whose proofs use it.

The effect of the AXIOM OF INDESCRIBABILITY seems to be to make the universe very large indeed. It can therefore be regarded as a strong infinity axiom. Indeed in its presence we can prove both of the axioms we stated in §3.1 as well as the AXIOM OF REPLACEMENT

†Proposition 3.4.1 (*Power-set principle*)
D *is a limit.*

Proof $\varnothing$ exists and so $\varnothing \in \mathbf{D}$ [AXIOM OF INDESCRIBABILITY], i.e. $\mathbf{D} \neq \varnothing$. Now let D_a be the earliest day to which a belongs. Then $(\forall a \in \mathbf{D})$ (D_a exists). Hence [AXIOM OF INDESCRIBABILITY] $(\forall a \in \mathbf{D})(a \in D_a \in \mathbf{D})$. So **D** is a limit day [proposition 1.5.7]. □

†Proposition 3.4.2 (*Infinity principle*)
There is a limit day earlier than **D**.

Proof Let D be the earliest limit to which $\varnothing$ belongs. Now $\varnothing \in \mathbf{D}$ and **D** is a limit [proposition 3.4.1]. Therefore D exists. Consequently $D \in \mathbf{D}$ [AXIOM OF INDESCRIBABILITY]; moreover D is evidently a non-empty limit. □

* The material in this section will not be used in the remainder of the book and can therefore be omitted.

†Proposition 3.4.3 (*Replacement principle*)

If $\tau(x)$ is a term in which the constant **D** *does not occur and $x_1, \ldots, x_n$ are the variables other than x on which it depends then for any $a \in \mathbf{D}$ this is a proposition*:

$$(\forall x_1, \ldots, x_n \in \mathbf{D})((\forall x \in a)(\tau(x) \in \mathbf{D}) \Rightarrow \{\tau(x) : x \in a\} \in \mathbf{D}).$$

Proof In the circumstances just described $\{\tau(x) : x \in a\}$ is a class [AXIOM OF SUBCOLLECTIONS] and hence a set [AXIOM OF INDESCRIBABILTY]. □

Indescribability The motivation behind the AXIOM OF INDESCRIBABILITY is apparently rather different from that generally given for the AXIOM OF REPLACEMENT, and it has a much greater effect on the size of the universe: its attraction for the lazy is that it makes the ceiling so high that only the most determined mountaineers are ever in any danger of bumping their heads on it.

4 Numbers

What we have done so far has been largely introductory in nature. The aim now is to show that the axioms of basic set theory are sufficient to ensure the existence in our theory of structures with the familiar properties we expect the sets of natural, rational, and real numbers to have.

4.1 Definition of natural numbers

We do not teach children the arithmetic of the natural numbers by first of all teaching them set theory: each of us has—we think—a complete understanding of what is meant by 'the natural numbers', and what it is to add and multiply them, before we read this chapter. So what we are engaged in here is a branch of applied mathematics. In §3.2 we gave a specification—the axioms of **PA**–of the properties we expect. If we are to do any worthwhile mathematics, it will certainly be necessary to have the natural numbers (or at any rate adequate proxies for them) available to us. This section and the next will be devoted to proving in **GA** the existence of a definite model of **PA**. (Whether the elements of this model are adequate proxies for the natural numbers is a question to which we shall return later.) We start by defining the notion of 'Dedekind algebra', which is intended to capture set-theoretically the content of those of the axioms of **PA** which implicitly define the successor operation. We make no mention of addition or multiplication yet: **GA** is sufficiently rich for them both to be definable in terms of the successor operation.

Definition

A *Dedekind algebra* is a successor algebra (A, f) such that $f: A \to A$ is injective and $A = \mathrm{Cl}_f(a)$ for some $a \in A - f[A]$.

Lemma 4.1.1

There exists a definite small Dedekind algebra.

Proof If D_ω is the earliest limit day, then D_ω is a set [axiom of infinity]. Moreover, we have already remarked in exercise 5 of §2.8 that the mapping $g: D_\omega \to D_\omega$ given by $g(x) = x \cup \{x\}$ is injective and $\varnothing \notin \mathrm{im}[g]$; so we can satisfy our requirements by letting $A = \mathrm{Cl}_g(\varnothing)$, $f = g|A$ and $a = \varnothing$. □

Definition

We let $(\boldsymbol{\omega}, s)$ denote a definite small Dedekind algebra. The elements of $\boldsymbol{\omega}$ are called *natural numbers*: the function s is called the *successor function* on $\boldsymbol{\omega}$.

More explicitly, this amounts to letting (say)

$$\mathfrak{D} = \iota x(x \text{ is a small Dedekind algebra})$$

and defining $\boldsymbol{\omega} = \text{dom}(\mathfrak{D})$ and $s = \text{im}(\mathfrak{D})$.

The use of abstraction terms This is the first time we have used the ι symbol to pick out an object which is not unique. The definition formalizes one given by Dedekind as follows (1888, No. 73): 'If in the consideration of a simply infinite system N ordered by a function ϕ [i.e. a Dedekind algebra (N, ϕ)] we entirely neglect the special character of the elements, simply retaining their distinctness and taking into account only the relations to one another in which they are placed by ϕ, then these elements are called *natural numbers* .... With reference to this freeing of the elements from every other content (abstraction) we are justified in calling numbers a free creation of the human mind.'

The systematic amnesia about the precise identity of $\boldsymbol{\omega}$ which our use of the ι symbol in this context represents is of course no more than a technical device. It is undoubtedly harmless, since it results in fewer things being provable about $\boldsymbol{\omega}$, not more; in particular, it avoids accidental properties of numbers (such as '$\varnothing \in 2$' or whatever) being provable in our theory. You may nevertheless feel unhappy with it. Frege was particularly scathing about such convenient forgetfulness (1903, p. 107) 'If, abstracting from the difference between my house and my neighbour's, I were to regard both houses as mine, the defect of my abstraction would soon be made clear.' If you share this view, replace the definition of $\boldsymbol{\omega}$ with an explicit one of your choice (such as the one implicit in the proof of lemma 4.1.1). Nothing whatsoever need then be changed in what follows.

Of course, if we intended to claim that the elements of $\boldsymbol{\omega}$ actually *are* the counting numbers zero, one, two, etc., then our forgetfulness about their exact identity would be more than a little disturbing. But that is not at all our intention. The most we could sensibly claim about $\boldsymbol{\omega}$ is that its elements have the same arithmetical properties as the counting numbers (although even that is open to some doubt: see the note 'Need a number-theorist believe Goodstein's theorem?' in §6.6) and that we may therefore treat them *as if* they were the counting numbers.

So the fact that there are many different possible candidates for $\boldsymbol{\omega}$ is a matter of complete indifference to us. The idea that the existence of different (but, as we shall show in corollary 4.1.5, necessarily isomorphic)

Dedekind algebras could seriously be regarded as an argument against realism in mathematics is quite bizarre. This has not deterred some philosophers from regarding it as one, though (e.g. Barker 1969).

It would make little difference in what follows if we chose instead to treat the notion of 'natural number' as primitive and added Peano's axioms as postulates. What we achieve by constructing what we need is no more than a certain economy of ontological commitment. Russell famously described the advantages of postulating what we need as those of 'theft over honest toil' (1919, p. 71). This is not quite right: it is rather that having already stolen enough to last us a lifetime we now have no further need of crime.

Definition

$0 = \iota! x(x \in \boldsymbol{\omega} - \mathrm{s}[\boldsymbol{\omega}])$('*zero*').

Justification By definition there is an element $0 \in \boldsymbol{\omega} - \mathrm{s}[\boldsymbol{\omega}]$ such that

$$\begin{aligned} \boldsymbol{\omega} &= \mathrm{Cl}_{\mathrm{s}}(0) \\ &= \{0\} \cup \mathrm{s}[\mathrm{Cl}_{\mathrm{s}}(0)] \text{ [proposition 2.3.1]} \\ &= \{0\} \cup \mathrm{s}[\boldsymbol{\omega}], \end{aligned}$$

and so if $0' \in \boldsymbol{\omega} - \mathrm{s}[\boldsymbol{\omega}]$ then $0 = 0'$. □

Definition

$1 = \mathrm{s}(0)$, $2 = \mathrm{s}(1)$, $3 = \mathrm{s}(2)$, $4 = \mathrm{s}(3)$.

It is a remarkable fact that all the arithmetical properties of the natural numbers can be derived from a small number of assumptions. In order to emphasize this point we restate the assumptions we are making about $\boldsymbol{\omega}$ in a slightly less compact notation.

Lemma 4.1.2

(*a*) $\mathrm{s}(m) = \mathrm{s}(n) \Rightarrow m = n$ *for all* $m, n \in \boldsymbol{\omega}$.
(*b*) $0 \neq \mathrm{s}(n)$ *for any* $n \in \boldsymbol{\omega}$.
(*c*) *If* $0 \in A \subseteq \boldsymbol{\omega}$ *and if* $n \in A \Rightarrow \mathrm{s}(n) \in A$, *then* $A = \boldsymbol{\omega}$.

Proof This is just a restatement of the definitions. □

Corollary 4.1.3 (*Simple principle of induction*)

If $\Phi(n)$ *is a formula such that* $\Phi(0)$ *and* $(\forall n \in \boldsymbol{\omega})(\Phi(n) \Rightarrow \Phi(\mathrm{s}(n)))$, *then* $(\forall n \in \boldsymbol{\omega})\Phi(n)$.

Proof Let $A = \{n \in \boldsymbol{\omega} : \Phi(n)\}$ and apply lemma 4.1.2(c). □

You are no doubt familiar with the method of defining a function g on the natural numbers by defining $g(0)$ and then giving an expression which defines $g(\mathrm{s}(n))$ in terms of $g(n)$. The justification for this method ('definition by recursion') is provided by the following theorem.

Theorem 4.1.4

If (A, f) is a successor algebra and $a \in A$, then there exists exactly one homomorphism $g : \boldsymbol{\omega} \to A$ such that $g(0) = a$.

Existence Let $g = \mathrm{Cl}_{s \times f}((0, a)) \subseteq \boldsymbol{\omega} \times A$ and

$$B = \{n \in \boldsymbol{\omega} : \text{there is exactly one } x \in A \text{ such that } n\ g\ x\}.$$

Then it is not hard to show that $0 \in B$ and that B is s-closed. Consequently $B = \boldsymbol{\omega}$ [lemma 4.1.2(c)], i.e. g is a function from $\boldsymbol{\omega}$ to A. It follows at once that it is the graph of the homomorphism we require (cf. exercise 1(a) of §2.8).

Uniqueness If $g, g' : \boldsymbol{\omega} \to A$ are homomorphisms such that $g(0) = g'(0)$ and if $C = \{n \in \boldsymbol{\omega} : g(n) = g'(n)\}$, then C is s-closed and $0 \in C$, so that $C = \boldsymbol{\omega}$ [lemma 4.1.2(c)] and hence $g = g'$. □

We are now in a position to show that the assumptions we made about the successor algebra $(\boldsymbol{\omega}, \mathrm{s})$ are sufficient to determine it completely (up to isomorphism).

Corollary 4.1.5

These two assertions are equivalent:
(i) (A, f) is a successor algebra isomorphic to $(\boldsymbol{\omega}, \mathrm{s})$;
(ii) (A, f) is a Dedekind algebra.

(i) ⇒ (ii) Immediate from the definition of $(\boldsymbol{\omega}, \mathrm{s})$.
(ii) ⇒ (i) If (A, f) is a Dedekind algebra, then the same argument which showed in the proof of theorem 4.1.4 that $\mathrm{Cl}_{s \times f}((0, a))$ is the graph of a homomorphism $\boldsymbol{\omega} \to A$ shows that its inverse is the graph of a homomorphism $A \to \boldsymbol{\omega}$. □

Definition

A family indexed by a subset of $\boldsymbol{\omega}$ is called a *sequence*.

Let us write $n + 1$ for $\mathrm{s}(n)$ and (if $n \neq 0$) $n - 1$ for the unique m such that $\mathrm{s}(m) = n$. (These are particular cases of notations we shall introduce in the next section.) With this terminology we can express theorem 4.1.4 in the following (more traditional) form.

Corollary 4.1.6 (*Simple recursion principle*)

If A is a collection, $f : A \to A$ is a mapping and $a \in A$, then there exists exactly one sequence (x_n) in A such that $x_0 = a$ and $x_{n+1} = f(x_n)$ for all $n \in \boldsymbol{\omega}$.

Proof This is just a rewording of theorem 4.1.4. □

Remark The sequence (x_n) referred to in this corollary is said to be obtained by 'applying simple recursion to (A, f) starting at a'. This method of defining sequences is of course widely used in mathematics: it is often signposted by the use of temporal terminology. For instance: 'Start by letting $x_0 = a$; then once x_n had been defined, let $x_{n+1} = f(x_n)$.'

Occasionally we may wish to define a sequence (x_n) in such a way that the definition of x_{n+1} in terms of x_n depends on the value of n. The simple recursion principle as we have stated it does not justify this, but an easy extension of it does.

Corollary 4.1.7 (*Simple principle of recursion with a parameter*)

If (f_n) is a sequence of functions from A to A and $a \in A$, then there exists a unique sequence (x_n) in A such that $x_0 = a$ and $x_{n+1} = f_n(x_n)$ for all $n \in \boldsymbol{\omega}$.

Proof If we define a mapping $h : \boldsymbol{\omega} \times A \to \boldsymbol{\omega} \times A$ by

$$h(n, x) = (n + 1, f_n(x)),$$

then it is straightforward to check that for any sequence (m_n, x_n) in $\boldsymbol{\omega} \times A$ these two assertions are equivalent:

(i) (m_n, x_n) is obtainable by simple recursion on the successor algebra $(\boldsymbol{\omega} \times A, h)$ starting at $(0, a)$;

(ii) $m_n = n$ for all $n \in \boldsymbol{\omega}$, $x_0 = a$ and $x_{n+1} = f_n(x_n)$ for all $n \in \boldsymbol{\omega}$.

The result follows. □

Example 1 If r is a relation on a collection A, let

$$r^0 = \mathrm{id}_A; \qquad r^{n+1} = r \circ r^n.$$

Then r^n is called the *nth iterate* of r on A. Observe that the zeroth iterate of r depends on A but that none of the others do.

Example 2 Suppose that $A \in \mathbf{D}$ and the define $f : \mathbf{D} \to \mathbf{D}$ by $f(X) = X \times A$. If we apply the simple recursion principle to the successor algebra $(\mathbf{D}, f)$ starting at A, we obtain a mapping $g : \boldsymbol{\omega} \to \mathbf{D}$ such that $g(0) = A$ and $g(n + 1) = g(n) \times A$ for all $n \in \boldsymbol{\omega}$. It is usual to denote $g(n - 1)$ by A^n. Hence $A^1 = A$ and $A^{n+1} = A^n \times A$ for all $n \geq 1$. Thus, for example, $A^4 = ((A \times A) \times A) \times A$. (For the sake of completeness we adopt the convention that $A^0 = \{\varnothing\}$.) The elements of A^n are called *n-tuples* in A.

Exercises

1. Show that there is no sequence (A_n) of collections such that $A_{n+1} \in A_n$ for all $n \in \omega$.
2. Fill in the details in the proof of theorem 4.1.4.
3. (a) If r and s are relations such that $r \circ s = s \circ r$, show that $r^n \circ s^m = s^m \circ r^n$ for all $m, n \in \omega$ and that $(r \circ s)^n = r^n \circ s^n$ for all $n \in \omega$.
   (b) If r is a relation, show that $(r^m)^n = (r^n)^m$ for all $m, n \in \omega$.
4. Fill in the details in the proof of corollary 4.1.7.

4.2 Arithmetic

It is now a straightforward matter to define the familiar algebraic operation on ω recursively. For example:

Definition

The *addition* function from $\omega \times \omega$ to ω given by $(m, n) \mapsto m + n$ is defined by the equations

$$m + 0 = m \tag{4.2.1}$$

$$m + \mathrm{s}(n) = \mathrm{s}(m + n). \tag{4.2.2}$$

Justification Apply the simple recursion principle to the successor algebra $(\mathrm{Cl_s}(m), \mathrm{s})$ starting at m. □

Remark 1 Note that

$$\mathrm{s}(n) = \mathrm{s}(n + 0) = n + \mathrm{s}(0) = n + 1$$

(thus coinciding, as promised, with the notation we introduced in §4.1), and so

$$m + (n + 1) = (m + n) + 1. \tag{4.2.2'}$$

All the familiar properties of addition of natural numbers follow from (4.2.1) and (4.2.2).

Example 1 It is reassuring to discover that

$$2 + 2 = 2 + (1 + 1) = (2 + 1) + 1 = 3 + 1 = 4.$$

Example 2 $k + (m + n) = (k + m) + n$ for all $k, m, n \in \omega$. We shall prove this by induction on n, noting first that

$$k + (m + 0) = k + m = (k + m) + 0 \quad [(4.2.1)],$$

so that it is true for 0. And if it is true for n then

$$
\begin{aligned}
k + (m + (n + 1)) &= k + ((m + n) + 1) && [(4.2.2)] \\
&= (k + (m + n)) + 1 && [(4.2.2)] \\
&= ((k + m) + n) + 1 && \text{by the induction hypothesis} \\
&= (k + m) + (n + 1) && [(4.2.2)],
\end{aligned}
$$

so that it is also true for $n + 1$. This completes the proof.

Remark 2 In the notation we introduced in example 1 of §4.1 our definition of addition amounts just to saying that

$$m + n = \mathrm{s}^n(\mathrm{s}^m(0)).$$

Of course, this equation would look more natural if we wrote functions on the right of their arguments. (Recall remark 2 of §2.2.)

Definition

The *multiplication* function from $\omega \times \omega$ to ω given by $(m, n) \mapsto mn$ is defined by the equations

$$m0 = 0 \tag{4.2.3}$$

$$m(n + 1) = mn + m. \tag{4.2.4}$$

Justification Apply the simple recursion principle to the algebra $(\omega, n \mapsto n + m)$ starting at 0. □

Remark 3 Equivalently, we could have defined $mn = (\mathrm{s}^m)^n(0)$.

Definition

The *exponentiation* function given by $(m, n) \mapsto m^n$ is defined by the equations

$$m^0 = 1 \tag{4.2.5}$$

$$m^{n+1} = m^n m \tag{4.2.6}$$

Justification Apply the simple recursion principle to $(\omega, n \mapsto nm)$ starting at 1. □

From these recursive definitions it is straightforward to derive all the elementary properties of the arithmetical operations. We shall not go into the details of any of the derivations here since they are largely a matter of routine: we shall, however, feel free to make use of any such properties when we need them.

Example 3 The formula $(\exists r \in \omega)(n = mr)$ is written $m|n$ and read 'm divides n'. It is easy to show that the divisibility relation which this formula defines on ω is a partial ordering. Somewhat confusingly, the greatest element of ω with respect to this partial ordering is 0 and the least element is 1. Indeed, the divisibility relation is contained in the partial ordering obtained from the usual one by swapping 0 and 1. If m and n are natural numbers, then $\{m, n\}$ has a supremum which is called their *least common multiple* and denote $\operatorname{lcm}(m, n)$; $\{m, n\}$ also has an infimum which is called their *greatest common divisor* and denoted $\gcd(m, n)$. Let us show now that for any $m, n \in \omega$ we have

$$mn = \operatorname{lcm}(m, n)\gcd(m, n). \tag{4.2.7}$$

If $\gcd(m, n) = 0$, then $m = n = 0$ and the result is trivial. So suppose not. Then certainly $\gcd(m, n)|n$, and so

$$m = \frac{m}{\gcd(m, n)}\gcd(m, n) \left| \frac{mn}{\gcd(m, n)} \right. .$$

But equally

$$n \left| \frac{mn}{\gcd(m, n)} \right. ,$$

whence

$$\operatorname{lcm}(m, n) \left| \frac{mn}{\gcd(m, n)} \right. .$$

Now there exist $r, s \in \omega$ such that

$$\operatorname{lcm}(m, n) = mr = ns.$$

Hence

$$\frac{m}{s} = \frac{n}{r}.$$

But

$$\frac{m}{s} \Big| m \quad \text{and} \quad \frac{n}{r} \Big| n, \quad \text{so} \quad \frac{m}{s} \Big| \gcd(m, n),$$

whence

$$\frac{mn}{\gcd(m, n)} = \frac{ns}{\gcd(m, n)} \frac{m}{s} \Big| ns = \operatorname{lcm}(m, n).$$

Equation (4.2.7) now follows.

The natural numbers The elementary arithmetic of the natural numbers has of course been known for thousands of years, but the principal features of the presentation we have given here did not emerge until much more recently. Euclid (*c*.300 BC) goes in some detail into elementary number theory (divisibility, prime numbers, etc.) but nowhere does he make use of mathematical induction. Indeed induction seems hardly to have been used at all as a method of proof until the mid-seventeenth century (Pascal in 1654, Fermat in 1659, Jakob Bernoulli in 1686).

The deduction of the elementary properties of the arithmetical operations from their recursive definitions was first done by Grassmann in 1861 and by Peirce in 1881. However, neither author made explicit the assumptions involved: a rigorous presentation of the theory first appeared in Dedekind's *Was sind und was sollen die Zahlen*? (1888)*. What is striking about this work is its modernity and perspicuity. For example, Dedekind was aware that the principle of induction does not in itself justify the method of definition by recursion and gave an explicit caution (later ignored by some other writers) to that effect. Dedekind's treatment has scarcely been bettered since and I have chosen to follow it quite closely here.

Exercises

1. Prove the following results by induction on n:
   (a) $1 + n = n + 1$.
   (b) $m + n = n + m$.
2. (a) If $f : A \to A$ is injective [respectively surjective], prove that the nth iterate $f^n : A \to A$ is injective [respectively surjective] for all $n \in \omega$.
   (b) Prove that $k + n = k + m \Rightarrow n = m$.

4.3 The ordering of the natural numbers

Lemma 4.3.1

s^{t} *is a strict partial ordering on* ω *and* s^{T} *is the associated partial ordering.*

Proof The only part which is not staringly obvious is showing that s^{t} is irreflexive. Now certainly

$$\begin{aligned} 0 \,\mathrm{s}^{\mathrm{t}}\, 0 &\Rightarrow 0 \in \mathrm{s}[\mathrm{Cl}_{\mathrm{s}}(0)] \quad \text{[proposition 2.3.3(b)]} \\ &= \mathrm{s}[\omega] \\ &\Rightarrow \text{contradiction} \quad \text{[lemma 4.1.2(b)].} \end{aligned}$$

And

$$\begin{aligned} \mathrm{s}(n) \,\mathrm{s}^{\mathrm{t}}\, \mathrm{s}(n) &\Rightarrow \mathrm{s}(n) \in \mathrm{s}[\mathrm{Cl}_{\mathrm{s}}(\mathrm{s}(n))] \quad \text{[proposition 2.3.3(b)]} \\ &\Rightarrow n \in \mathrm{Cl}_{\mathrm{s}}(\mathrm{s}(n)) \quad \text{[lemma 4.1.2(a)]} \\ &= \mathrm{s}[\mathrm{Cl}_{\mathrm{s}}(n)] \quad \text{[proposition 2.3.1(b)]} \\ &\Rightarrow n \,\mathrm{s}^{\mathrm{t}}\, n \quad \text{[proposition 2.3.3(b)].} \end{aligned}$$

It follows by induction that s^{t} is an irreflexive relation on ω. □

* Some material suggestive of Dedekind's development is to be found in Part III of Frege's *Begriffschrift* (1879), but Frege does not explicitly define the natural numbers here. An early draft of Dedekind's work, entitled *Gedanken über die Zahlen* (which he wrote between 1872 and 1878), is printed in Dugac (1976, appendice LVI).

So s^t is a strict partial ordering on ω containing s. In fact it is the *only* one (see exercise 4 below). So if we want to have $n < s(n)$ for all $n \in \omega$ then we have no choice but to make the following definition.

Definition

We write $<$ and $\leq$ for s^t and s^T respectively. Whenever we refer to ω as a [strictly] partially ordered set it will be with respect to this [strict] partial ordering.

Theorem 4.3.2 (*Least element principle*)

Every non-empty subset of ω has a least element.

Proof Suppose that B is a subset of ω which does not have a least element. Now

$$\omega = \mathrm{Cl}_s(0) = s^T[0] \quad \text{[proposition 2.3.3(a)]}$$

and so 0 is certainly a lower bound for B. And if n is a lower bound for B, then $n \notin B$ (since otherwise n would be the least element of B, contrary to hypothesis), so that

$$B \subseteq s^t[n] = s^T[s(n)] \quad \text{[proposition 2.3.2(b)]}$$

and hence $s(n)$ is a lower bound for B. Consequently by induction every element of ω is a lower bound for B. So if $n \in B$ then in particular $s(n)$ is a lower bound for B and so $s(n) \leq n$, contradicting the definition of the partial ordering. □

Corollary 4.3.3

$\leq$ is a total order relation on ω.

Proof If $m, n \in \omega$, then $\{m, n\}$ has a least element [theorem 4.3.2]; this least element is either m, in which case $m \leq n$, or n, in which case $n \leq m$. □

Exercises

1. (a) Show that $m < n \Leftrightarrow s(m) \leq n$ for all $m, n \in \omega$.
   (b) Prove by induction on n that $q < r \Rightarrow n + q < n + r$ for all $n, q, r \in \omega$.
2. Prove that if $m \leq n$ then there exists a unique element $n - m \in \omega$ such that $n = m + (n - m)$. [*Existence* Consider the least element of $\{r \in \omega : m + r \geq n\}$.]
3. Let n be a natural number.
   (a) Prove that $n + 1$ is the unique successor of n.
   (b) Show that if $n \neq 0$ then $n - 1$ is the unique predecessor of n.
4. Show that $<$ is the only strict partial ordering on ω with the property that $n < s(n)$ for all $n \in \omega$.

Definition

$\boldsymbol{n} = \{m \in \omega : m < n\}$.

Exercise

5. Show that the proper initial subsets of ω are precisely the sets $\boldsymbol{n}$ for $n \in \omega$.

Corollary 4.3.4

If $A \subseteq \omega$ is such that $\mathbf{n} \subseteq A \Rightarrow n \in A$ for every $n \in \omega$, then $A = \omega$.

Proof Suppose $A \neq \omega$. So $\omega - A$ is non-empty and therefore [theorem 4.3.2] has a least element n. Hence $\mathbf{n} \subseteq A$ and so by hypothesis $n \in A$. Contradiction. □

Corollary 4.3.5 (*General principle of induction*)

If $\Phi(n)$ is a formula such that

$$(\forall n \in \omega)((\forall m < n)\Phi(m) \Rightarrow \Phi(n)),$$

then $(\forall n \in \omega)\Phi(n)$.

Proof Let $A = \{n \in \omega : \Phi(n)\}$ and apply corollary 4.3.4. □

Theorem 4.3.6 (*General recursion principle with a parameter*)

If for each $n \in \omega$ we have a mapping $s_n : \mathfrak{P}(A) \to A$, then there exists exactly one mapping $g : \omega \to A$ such that $g(n) = s_n(g[\mathbf{n}])$ for all $n \in \omega$.

Proof Exercise. □

Remark 1 The use of the general recursion principle in the course of a proof is often signposted by temporal terminology such as: 'Once $g(r)$ has been defined for all $r < n$, let $g(n) = s_n(g[\mathbf{n}])$.'

Proposition 4.3.7

Suppose that r is a relation on A.

(a) $r^{\mathrm{t}} = \bigcup_{n \geq 1} r^n$.

(b) $r^{\mathrm{T}} = \bigcup_{n \geq 0} r^n$.

Proof (a) It is easy to show by induction on n that $r^n \subseteq r^{\mathrm{t}}$ for all $n \geq 1$; consequently $\bigcup_{n \geq 1} r^n \subseteq r^{\mathrm{t}}$. And if $x\, r^n\, y$ and $y\, r^m\, z$ then $x\, r^{n+m}\, z$, from which it follows that $\bigcup_{n \geq 1} r^n$ is transitive, so that $r^{\mathrm{t}} \subseteq \bigcup_{n \geq 1} r^n$. Thus $r^{\mathrm{t}} = \bigcup_{n \geq 1} r^n$.

(b) $r^{\mathrm{T}} = r^{\mathrm{t}} \cup \mathrm{id}_A$ [proposition 2.3.2(a)]

$= r^0 \cup \bigcup_{n \geq 1} r^n$ [(a)]

$= \bigcup_{n \geq 0} r^n$. □

Definition

A sequence indexed by $\mathbf{n}$ is called a *string*; the natural number n (which is evidently unique) is called the *length* of the string. The collection $\bigcup_{n \in \omega} {}^{\mathbf{n}}A$ of all strings in A is denoted String(A).

Exercises

6. Prove the general recursion principle.
7. (a) If (A_n) is a sequence of sets, show that there is a sequence $(\bigtimes_{r<n} A_r)$ of sets such that $\bigtimes_{r<1} A_r = A_0$ and $\bigtimes_{r<n+1} A_r = (\bigtimes_{r<n} A_r) \times A_n$ for all $n \in \boldsymbol{\omega}$.
   (b) If A is a set, show that $\bigtimes_{r<n} A = A^n$ for all $n \in \boldsymbol{\omega}$.
   (c) Show that if $(A_r)_{r \in n}$ is a sequence of sets then the mapping $\Pi_{r \in n} A_r \to \bigtimes_{r<n} A_r$ given by $f \mapsto (f(0), f(1), \ldots, f(n-1))$ is bijective. (It is common to identify $\Pi_{r \in n} A_r$ with $\bigtimes_{r<n} A_r$ by means of this one-to-one correspondence—thus extending the scope of the identification introduced in remark 8 of §2.2—and in particular to identify the collection ${}^nA$ of strings of length n in A with the collection A^n of n-tuples in A.)

Remark 2 When $n = 1$ the identification procedure we have just referred to involves identifying an element a of a set A with the function $\{(0, a)\}$ from $\mathbf{1}$ to A. We can therefore rewrite the simple recursion principle in the following (more abstract) form: $(\boldsymbol{\omega}, \mathrm{s})$ is a successor algebra and $0 : \mathbf{1} \to \boldsymbol{\omega}$ is a mapping such that if (A, f) is a successor algebra and $a : \mathbf{1} \to A$ is a mapping, then there is exactly one homomorphism $g : \boldsymbol{\omega} \to A$ which makes the diagram

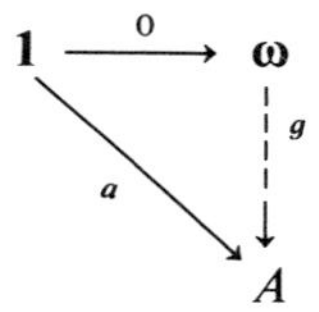

commute.

Exercises

8. Show that the property given in remark 2 above characterizes the successor algebra $(\boldsymbol{\omega}, \mathrm{s})$ up to isomorphism.
9. If r is a relation on A, show that $(r \cup \mathrm{id}_A)^n = \bigcup_{m=0}^{n} r^m$ for all $n \in \boldsymbol{\omega}$.

4.4 Characterization of the ordering of the natural numbers

We intend now to characterize the order structure of the set of natural numbers. First we must introduce two properties which help us to describe the *size* of a collection: this is a theme we shall return to in Chapter 5.

Definition

A collection A is said to be *countable* if either $A = \varnothing$ or there exists a sequence whose range is A; A is *finite* if there exists a string whose range is A.

The set of all finite subcollections of A is denoted $\mathfrak{F}(A)$. We let $\mathfrak{F}(A, B)$ denote the collection of all finite partial functions from A to B, i.e.

functions f such that $\mathrm{dom}[f]$ is a finite subcollection of A and $\mathrm{im}[f]$ is a subcollection of B.

heorem 4.4.1

Every non-empty finite partially ordered collection has a maximal element.

roof Let $P(n)$ be the assertion 'If there exists a string of length n whose range is a partially ordered collection, then that collection has a maximal element.' Then $P(1)$ is obviously true. So suppose now that $P(n)$ is true. If $(A, \leq)$ is a partially ordered collection such that $A = \{x_r : r < n + 1\}$, then by our induction hypothesis $\{x_r : r < n\}$ has a maximal element y: if $y \leq x_n$ then x_n is a maximal element of A, and if $y \nleq x_n$, then y is a maximal element of A. Hence $P(n + 1)$ is true. It follows by induction that $P(n)$ is true for all $n \in \omega - \{0\}$, which proves the theorem. □

roposition 4.4.2 *There exist definite bijective mappings as follows*:

(*a*) $\omega \times \omega \to \omega$;
(*b*) $\omega \times \{0, 1\} \to \omega$;
(*c*) $\boldsymbol{m} \cup \boldsymbol{n} \to \{r \in \omega : r < m + n\}$;
(*d*) $\boldsymbol{m} \times \boldsymbol{n} \to \{r \in \omega : r < mn\}$;
(*e*) ${}^{n}\boldsymbol{m} \to \{r \in \omega : r < m^n\}$;
(*f*) $\mathfrak{F}(\omega) \to \omega$.

roof The mappings determined by the following functions will do:
(a) $(m, n) \mapsto 2^n(2m + 1) - 1$.
(b) $(n, i) \mapsto 2n + i$.
(c) $(r, i) \mapsto im + r$.
(d) $(r, s) \mapsto rn + s$.
(e) $(n_r)_{r \in \boldsymbol{n}} \mapsto \Sigma_{r \in \boldsymbol{n}} n_r m^r$.
(f) $A \mapsto \Sigma_{n \in A} 2^n$.
The proofs that these mappings are bijective are largely a matter of elementary arithmetic and we therefore omit them. □

xample 1 Evidently $\boldsymbol{n}$ is finite for all $n \in \omega$; in particular, $\varnothing$ is finite.

xample 2 Every finite collection is countable, whereas ω is countable but not finite (since if it were finite it would have a greatest element [theorem 4.4.1]).

xample 3 Every subcollection of a finite [respectively countable] collection is finite [respectively countable]. Hence $\mathfrak{F}(A) = \mathfrak{B}(A)$ iff A is finite. If $f : A \twoheadrightarrow B$ is surjective and A is finite [respectively countable], then B is finite [respectively countable].

xample 4 $\mathfrak{F}(\omega)$ is countable [proposition 4.4.2(f)].

xample 5 $\omega \times \omega$ is countable [proposition 4.4.2(a)]. Hence by an easy induction ω^n is countable for all $n \in \omega$.

Example 6 Suppose for a moment that $\mathfrak{P}(\omega)$ is countable. So there is a surjective mappin $f:\omega \to \mathfrak{P}(\omega)$: if $A = \{n \in \omega : n \notin f(n)\}$, then there exists $n_0 \in \omega$ such that $f(n_0) =$ (since f is surjective), so that

$$n_0 \in A \Leftrightarrow n_0 \notin f(n_0) \Leftrightarrow n_0 \notin A,$$

which is absurd. Therefore $\mathfrak{P}(\omega)$ is not countable.

Remark 1 Our proof that $\mathfrak{P}(\omega)$ is uncountable is irredeemably impredicative sinc the definition of the set A depends on a set (the function f) whose birthda must be *after* that of A.

Exercises

1. Show that a collection is finite iff it is equinumerous with $\boldsymbol{n}$ for some $n \in \omega$.
2. Suppose that $(A, \leq)$ is a totally ordered collection. Show that any sequenc $(a_n)_{n \in \omega}$ in A has a monotonic subsequence. (We say that $(a_{n_r})_{r \in \omega}$ is a *subsequen* of (a_n) if (n_r) is a strictly increasing sequence in ω.) [Let

$$B = \{n \in \omega : a_n < a_r \text{ for all } r > n\}$$

and consider two cases according to whether or not B is finite.]
3. (a) If r is the relation on $\mathfrak{P}(A)$ defined between X and Y by the formula 'the exists $a \in A$ such that $Y = X \cup \{a\}$', show that $\mathfrak{F}(A) = \mathrm{Cl}_r(\varnothing)$.
   (b) Show that $X, Y \in \mathfrak{F}(A) \Rightarrow X \cup Y \in \mathfrak{F}(A)$.
   (c) Show that if A is finite then $\mathfrak{P}(A)$ is finite.
4. Let r be a relation on A.
   (a) Show that these four assertions are equivalent:
      (i) there exists a strict partial ordering on A containing r;
      (ii) r^t is a strict partial ordering on A;
      (iii) r^t is irreflexive;
      (iv) there do not exist $x_0, \ldots, x_{n-1} \in A$ such that $x_0 \, r \, x_1 \, r \ldots r \, x_{n-1} \, r \, x_0$.
   We say that r is a *strict subordering* if it satisfies these conditions.
   (b) r is called a *predecessor* relation on A if it is a strict subordering and i addition:

   $$\text{if } x_0 \, r \, x_1 \, r \ldots r \, x_{n-1} \quad \text{and} \quad x_0 \, r \, x_{n-1}, \quad \text{then } n = 2.$$

   Show that if r is a predecessor relation on A then $x \, r \, y$ iff x is a predecessor of with respect to the strict partial ordering r^t.
   (c) If $<$ is a strict partial ordering on A, show that the formula 'x is predecessor of y' defines a predecessor relation $(<)$ on A contained in $<$
   (d) If A is finite, show that $(<)^t$ is equal to $<$.
   (e) Give an example where $(<)^t$ is a proper subcollection of $<$.

Remark 2 The preceding exercise shows that a partial ordering on a finite collectio is determined by its predecessor relation. It is common to illustrate finit partially ordered collections by pictures in which a line from a poir labelled x to a higher point labelled y indicates that x is a predecessor c y. Such pictures are called *Hasse diagrams*.

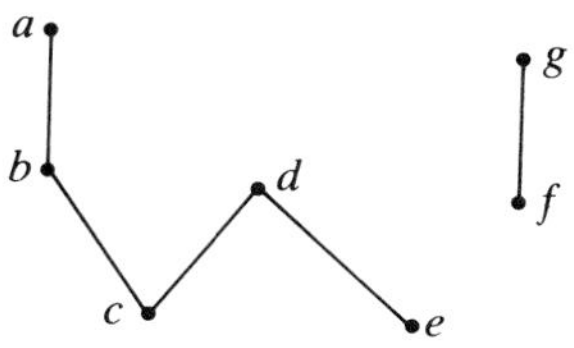

Figure 2

xercise 5

5. List the elements of the partial ordering on the collection $\{a, b, c, d, e, f, g\}$ which has the Hasse diagram illustrated in Fig. 2.

We have already characterized the structure of the successor algebra (ω, s) [corollary 4.1.5]; what we shall do now is characterize the structure of the partially ordered set $(\boldsymbol{\omega}, \leq)$.

heorem 4.4.3

These two assertions are equivalent:

(i) $(A, \leq)$ *is isomorphic to* $(\boldsymbol{\omega}, \leq)$;

(ii) $(A, \leq)$ *is a non-finite totally ordered collection every proper initial subcollection of which is finite.*

(i) $\Rightarrow$ (ii) We have already shown that $(\boldsymbol{\omega}, \leq)$ is a totally ordered collection and that it is not finite. Moreover, if B is a proper initial subcollection of $\boldsymbol{\omega}$ then there exists $n \in \boldsymbol{\omega} - B$, so that $B \subseteq \boldsymbol{n}$ and hence B is finite. The result follows.

(ii) $\Rightarrow$ (i) Suppose that $(A, \leq)$ has the properties described in (ii). Let us first observe that if a subcollection B of A is non-empty, then it has a least element, since if $b \in B$ then either $\{x \in B : x < b\}$ is empty, in which case b is the least element of B, or it is finite and non-empty, in which case it has a least element [theorem 4.4.1] which is evidently also the least element of B. Now by the general recursion principle we can define a mapping $f : \boldsymbol{\omega} \to A$ as follows: when $f(m)$ has been defined for all $m < n$, let $f(n)$ be the least element of $A - f[\boldsymbol{n}]$ (which is non-empty since A is non-finite). Now $f : \boldsymbol{\omega} \to A$ is evidently strictly increasing; moreover, it is surjective, because $f[\boldsymbol{\omega}]$ is an initial subcollection of A which is not finite and must therefore be A itself. Since $\boldsymbol{\omega}$ is totally ordered, it follows that $f : \boldsymbol{\omega} \to A$ is an isomorphism. □

xercises

6. Show that a partially ordered collection is directed iff every finite subcollection has an upper bound.
7. Show that if $(A, \leq)$ is a partially ordered collection, then these three assertions are equivalent:

(i) $(A, \leq)$ is isomorphic to $(\boldsymbol{n}, \leq)$ for some $n \in \omega$;
(ii) A is finite and $\leq$ is a total ordering;
(iii) Every non-empty subcollection of A has a greatest and a least element.

8. Show that A is finite iff there exists a function f on A such that the only f-close subcollections of A are $\varnothing$ and A itself.

4.5 The rational line

The rational numbers as we know them from school have the followin properties (among others):

Between any two rational numbers there is another rational number
Greater than any rational number there is another rational number;
Less than any rational number there is another rational number;
There are only countably many rational numbers in all.

We intend to show in this section that there is a definite ordered set i our theory which has the formal analogues of these four properties an moreover, that it is characterized uniquely (up to isomorphism) by them.

Definition

A subcollection B of an ordered collection $(A, \leq)$ is said to be *strict dense* in A if for all $x, y \in A$ such that $x < y$ there exists $z \in B$ such th $x < z < y$; if A is strictly dense in itself, then we may say just that it *strictly dense.*

Remark 1 It amounts to the same to say that $(A, \leq)$ is strictly dense if no eleme of A has a predecessor or a successor. If this is the case then A cannot b finite (unless it is trivial).

Remark 2 Equivalently, B is strictly dense in the ordered collection $(A, \leq)$ iff $a =$ $\sup\{x \in B : x < a\}$ for all $a \in A$.

Lemma 4.5.1

There exists a definite countable strictly dense ordered set with no greate or least element.

Proof Let us define a relation $\leq$ on $\mathfrak{F}(\omega) - \{\varnothing\}$ by writing $A \leq B$ iff eithe $A = B$ or the least element in one of A and B but not in both is in A. is easy to verify that $(\mathfrak{F}(\omega) - \{\varnothing\}, \leq)$ is a strictly dense ordered set wit no greatest or least element. And as we have already observed in examp 4 of §4.4, $\mathfrak{F}(\omega)$ (and hence $\mathfrak{F}(\omega) - \{\varnothing\}$) is countable. □

Definition

The ordered set described in lemma 4.5.1 is denoted $(\mathbf{Q}, \leq)$ and called th *rational line.*

Warning The specification we have just given of the rational line is not a specification of the set of rational *numbers* since it omits any reference to their algebraic (ordered field) structure: indeed their ordering is not sufficient to determine this structure uniquely. The advantage of this eccentric approach is that it avoids a somewhat lengthy detour into a fairly routine piece of algebra. Constructions of the rational numbers which include their algebraic structure may be found in many textbooks.

The properties we have used to define the ordered set $(\mathbf{Q}, \leq)$ are sufficient to determine it up to isomorphism.

heorem 4.5.2 (Cantor 1895)
These two assertions are equivalent:
(i) $(A, \leq)$ is isomorphic to $(\mathbf{Q}, \leq)$;
(ii) $(A, \leq)$ is a non-empty countable strictly dense ordered set with no greatest or least element.

(i) ⇒ (ii) Trivial.
(ii) ⇒ (i) Let (a_n) and (b_n) be sequences whose images are A and $\mathbf{Q}$ respectively. We construct a mapping $f : A \to \mathbf{Q}$ recursively: once $f(a_r)$ has been defined for all $r < m$ let $f(a_m)$ be the element b_n of least possible index such that it bears the same order relation to $f(a_0), \ldots, f(a_{m-1})$ as a_m bears to $a_0, \ldots, a_{m-1}$. (Such an element always exists because the ordering on $\mathbf{Q}$ is strictly dense and has no greatest or least element.) Now it is clear that the resulting mapping $f : A \to \mathbf{Q}$ is strictly increasing (and therefore injective); we will be finished if we can show that it is surjective. So suppose not. There therefore exists an element $b_n \in \mathbf{Q} - f[A]$ of least possible index. If $r < n$ then $b_r \in f[A]$ and so we can let m_r be the least natural number such that $f(a_{m_r}) = b_r$. If m is the least natural number greater than $m_0, \ldots, m_{n-1}$ such that a_m bears the same order relation to $a_{m_0}, \ldots, a_{m_{n-1}}$ as b_n bears to $f(a_{m_0}), \ldots, f(a_{m_{n-1}})$, then it is clear that $f(a_m) = b_n$. Contradiction. □

xercise Let $(A, \leq)$ be a countable partially ordered collection.
(a) Prove that there exists an injective increasing mapping of $(A, \leq)$ into $(\mathbf{Q}, \leq)$.
(b) Deduce that the partial ordering of A is contained in a total ordering.

.6 The real line

We now turn to the construction of the real line. The key property here, on which all undergraduate courses of real analysis are based, is that every non-empty set of real numbers which has an upper bound has a least upper bound: put more briefly, the real line is boundedly complete.

Lemma 4.6.1

There exists a definite non-empty boundedly complete ordered set with n greatest or least element which has a definite countable strictly dense subset

Justification Let $\mathscr{A}$ be the set of all non-empty proper initial subsets c $\mathbf{Q}$ which have no greatest element. $\mathscr{A}$ is evidently totally ordered b inclusion and has no smallest or largest element. Moreover, if $\mathscr{B}$ is a chai in $\mathscr{A}$, all the elements of which are contained in some element of $\mathscr{A}$, the it is easy to verify that $\bigcup \mathscr{B} \in \mathscr{A}$; it follows that $\mathscr{A}$ is boundedly complet Finally $\{\downarrow x - \{x\} : x \in \mathbf{Q}\}$ is a countable subset of $\mathscr{A}$ since $\mathbf{Q}$ is countabl and is strictly dense in $\mathscr{A}$ since if $A, B \in \mathscr{A}$ and $A \subset B$ then there exist $x \in B - A$ and so $A \subset \downarrow x - \{x\} \subset B$. □

Definition

The ordered set described in lemma 4.6.1 is denoted $(\mathbf{R}, \leq)$ and called th *real line* or the *continuum*. The definite countable strictly dense subset c $\mathbf{R}$ referred to in the definition is evidently isomorphic to $\mathbf{Q}$ [theorer 4.5.2]: we shall identify it with $\mathbf{Q}$ in what follows.

Proposition 4.6.2

If $(A, \leq)$ is a boundedly complete ordered collection and $f : \mathbf{Q} \to A$ is normal mapping, then there exists exactly one normal mapping $\bar{f} : \mathbf{R} \to$ which makes the diagram

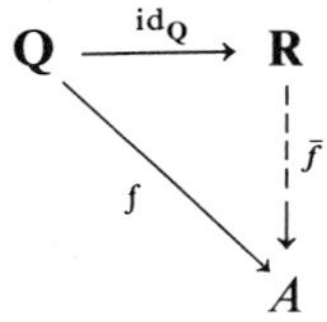

commute (i.e. agrees with f on $\mathbf{Q}$).

Uniqueness If $a \in \mathbf{R}$, then in order that $\bar{f}$ should be normal we must have

$$\bar{f}(a) = \sup_{x \in \mathbf{Q}, x<a} \bar{f}(x) = \sup_{x \in \mathbf{Q}, x<a} f(x),$$

whence the uniqueness of $\bar{f}$.

Existence What we have just said suggests that we should define

$$\bar{f}(a) = \sup_{x \in \mathbf{Q}, x<a} \bar{f}(x).$$

Then certainly if $a \in \mathbf{Q}$ we have $\bar{f}(a) = f(a)$ since f is normal. And if C a subset of $\mathbf{R}$ which has a supremum c, then

$$\sup \bar{f}[C] = \sup_{x \in C} \sup_{y \in \mathbf{Q}, y<x} f(y) = \sup_{y \in \mathbf{Q}, y<c} f(y) = \bar{f}(c).$$

This shows that $\bar{f}$ is normal. □

Remark 1 If we had chosen to adopt the conventional approach and include in our specification of **Q** its familiar ordered field structure rather than merely an order structure, it would now be a straightforward (albeit extremely tedious) matter to show by using proposition 4.6.2 that this extends uniquely to an ordered field structure on **R**. Note that by omitting to do this we have failed to provide a foundation for the traditional treatment of elementary analysis since it is evidently essential in mathematics to be able to add real numbers together. However, we shall not need the algebraic structure of the real numbers in this book (except in some of the exercises). If you are interested, get the details of the construction from another textbook.

Remark 2 The completion $\bar{\mathbf{R}}$ of **R** is isomorphic to the ordered set obtained from **R** by adjoining top and bottom elements (cf. exercise 4 of §2.7); it is usual to denote these elements $+\infty$ and $-\infty$ respectively and to call $\bar{\mathbf{R}}$ the *extended real line.*

The properties we have used to define $(\mathbf{R}, \leq)$ are sufficient to determine it up to isomorphism.

Theorem 4.6.3

These two assertions are equivalent:

(i) $(A, \leq)$ *is isomorphic to* $(\mathbf{R}, \leq)$;

(ii) $(A, \leq)$ *is a non-empty boundedly complete ordered collection with no greatest or least element which has a countable strictly dense subcollection.*

(i) ⇒ (ii) Trivial.

(ii) ⇒ (i) If B is a countable strictly dense subcollection of A then there is an isomorphism $f : \mathbf{Q} \to B$ [theorem 4.5.2]. This isomorphism is certainly strictly normal and therefore has a normal extension $\bar{f} : \mathbf{R} \to A$ [proposition 4.6.2]. Now if $a \in A$ and we let $C = f^{-1}[\{x \in B : x < a\}]$ then

$$\begin{aligned} \bar{f}(\sup C) &= \sup \bar{f}[C] \\ &= \sup f[C] \\ &= \sup\{x \in B : x < a\} \\ &= a. \end{aligned}$$

So $\bar{f}: \mathbf{R} \to A$ is surjective. Moreover, it is evidently strictly increasing. Hence (since $\mathbf{R}$ is totally ordered) it is an isomorphism. □

Exercises

1. If $(A, \leq)$ is a totally ordered collection which has a countable strictly dense subcollection, show that it can be embedded in $(\mathbf{R}, \leq)$.
2. Give an example of a strictly dense ordered set $(A, \leq)$ without a greatest or a least element which is equinumerous with but not isomorphic to $(\mathbf{R}, \leq)$. [Remove one point from $\mathbf{R}$.]

5 Cardinals

In this chapter we shall study in some detail the concept of 'equinumerosity', which was introduced in §2.2 to express in our theory the idea of collections having the same size. In order to deal with this notion it will be helpful to associate with each collection an object which we can look on as *being* its size.

Now when the collections involved are finite, we already have available to us objects (the natural numbers) which achieve this: it is an intuitively familiar fact that every finite collection is equinumerous with $\boldsymbol{n}$ for exactly one natural number n (which is of course said to be the 'number' of elements in the collection). One of the things we shall do in this chapter is to show that this 'fact' is indeed provable in our theory.

But what if the collections involved are not finite? A naive conjecture might be that all such collections are equinumerous and that we therefore need only one object ('infinity') to measure their size. But this is not the case: $\boldsymbol{\omega}$ and $\mathfrak{P}(\boldsymbol{\omega})$ are not equinumerous, nor is either of them finite (example 6 of §4.4). And this is not an isolated example: the same argument shows that $\mathfrak{P}(\mathfrak{P}(\boldsymbol{\omega}))$, $\mathfrak{P}(\mathfrak{P}(\mathfrak{P}(\boldsymbol{\omega})))$, etc. all have different sizes. We are therefore forced to create a fairly elaborate theory of size (also known as cardinality) in order to have a mechanism for calculating the size of non-finite collections.

5.1 Mapping properties of collections

The following results—Bernstein's equinumerosity theorem and Cantor's theorem in particular—are fundamental to the theory of cardinals.

Lemma 5.1.1

If $A \subseteq B \subseteq C$ and A is equinumerous with C then B is equinumerous with both A and C.

Proof It is an easy exercise to check that if $f : C \to A$ is bijective then the mapping $g : C \to B$ given by

$$g(x) = \begin{cases} f(x) & \text{if } x \in \mathrm{Cl}_f(C - B) \\ x & \text{if } x \in C - \mathrm{Cl}_f(C - B) \end{cases}$$

is bijective. □

Theorem 5.1.2 (*Bernstein's equinumerosity theorem*)

If there exist injective mappings $f : A \to B$ and $g : B \to A$, then there exists a bijective mapping $A \to B$.

Proof $g[f[A]] \subseteq g[B] \subseteq A$ and $g \circ f : A \to g[f[A]]$ is bijective. So [lemma 5.1.1] there exists a bijective mapping $h : A \to g[B]$. But $g^{-1} : g[B] \to B$ is bijective. So $g^{-1} \circ h : A \to B$ is bijective. □

Bernstein's equinumerosity theorem This theorem, which was conjectured by Cantor (1883a, 1895), was proved by his pupil, Felix Bernstein in 1897 when he was barely 19; a slightly simplified version of his proof was published by Borel in an appendix to his *Leçons sur la théorie des fonctions* (1898). The version we have sketched here, which uses Dedekind's *Kettentheorie* to avoid mention of the natural numbers, was found by Dedekind following a conversation with Bernstein* and communicated to Cantor in a letter (1897). However, Dedekind never published it and it was rediscovered by Zermelo in 1906 (see Poincaré 1906; Peano 1906; Zermelo 1908*b*).

In Britain and Germany the result is usually called the 'Schröder/Bernstein theorem' because of an attempted proof by Schröder (1898) the fallacy in which was not exposed in print until Korselt (1911) (although it had been pointed out to Schröder and acknowledged by him in 1902). According to another tradition, common in France and Italy, the result is known as the 'Cantor/Bernstein theorem', perhaps in recognition of Cantor's statement of it or possibly on account of an ambiguous footnote in Borel's *Leçons* (1898, p. 105n.). In any event it is clear that Cantor never had a direct proof but could only deduce the theorem from a formulation of the cardinal comparability principle (discussed in §7.3), a result for which in turn he never obtained a convincing proof (see Cantor 1903).

Recall that ${}^{A}\mathbf{2}$ denotes the collection of all functions $A \to \mathbf{2}$; we write ${}^{(A)}\mathbf{2}$ to denote the subcollection consisting of those functions $f: A \to \mathbf{2}$ such that $f(x) = 0$ for all but finitely many $x \in A$. The *characteristic function* of a collection $B \subseteq A$ is the element c_B of ${}^{A}\mathbf{2}$ given by

$$c_B(x) = \begin{cases} 1 & \text{if } x \in B \\ 0 & \text{if } x \in A - B \end{cases}$$

Proposition 5.1.3

There exist definite bijective mappings as follows:

(*a*) ${}^{A}\mathbf{2} \to \mathfrak{P}(A)$

(*b*) ${}^{(A)}\mathbf{2} \to \mathfrak{F}(A)$.

* A manuscript copy of a very similar proof, apparently dated 11/7/1887, was found among Dedekind's papers (Dedekind 1932, Vol. III, pp. 447–8).

Proof In each case the function $f \mapsto f^{-1}[1]$ is a one-to-one correspondence; its inverse is given by $B \mapsto c_B$. □

Remark There are also definite bijective mappings $\mathfrak{P}(A) \to \mathfrak{P}(A, \mathbf{1})$ and $\mathfrak{F}(A) \to \mathfrak{F}(A, \mathbf{1})$ given by $B \mapsto B \times \mathbf{1}$; their inverses are given by $f \mapsto \text{dom}[f]$.

Theorem 5.1.4 (Cantor 1892)

There exist injective mappings $A \to \mathfrak{P}(A)$ but not surjective ones.

Proof $x \mapsto \{x\}$ clearly determines an injective mapping $A \to \mathfrak{P}(A)$. Suppose now that $f: A \to \mathfrak{P}(A)$ is a surjective mapping and let $B = \{x \in A : x \notin f(x)\}$. Then $B \in \mathfrak{P}(A)$ and so there exists $y \in A$ such that $B = f(y)$. Hence

$$y \in B \Leftrightarrow y \notin f(y) \Leftrightarrow y \notin B,$$

which is absurd. □

Exercises

1. Write out the details of the proof of lemma 5.1.1.
2. Does there exist a collection A such that $(\forall a)(a \subseteq A \Rightarrow a \in A)$?

5.2 Definition of cardinals

The results of §5.1 are a start, but they are difficult to manipulate as they stand. What we need to do is to index collections in such a way that collections are equinumerous iff they have the same index. To do this we use the cutting-down trick which we introduced in §1.5.

Definition

The collection $\langle X : X \text{ and } A \text{ are equinumerous}\rangle$ is called the *cardinal* of A and denoted $\text{card}(A)$.

What cardinal numbers are is less important than what they do.

Proposition 5.2.1

A and B are equinumerous iff $\text{card}(A) = \text{card}(B)$.

Necessity If A and B are equinumerous then the collections equinumerous with A are precisely those equinumerous with B, and so $\text{card}(A) = \text{card}(B)$.

Sufficiency Every collection A is equinumerous with itself and so $\text{card}(A) \neq \varnothing$ [proposition 1.5.4]. Consequently, if $\text{card}(A) = \text{card}(B)$ there must exist a collection X which belongs to both $\text{card}(A)$ and $\text{card}(B)$, and hence is equinumerous with both A and B: it follows that A and B must be equinumerous. □

The definition of cardinality The concept of cardinality does not occur explicitly in Cantor's early work on equinumerosity (1874) and when it does appear in 1878 it is only in compound phrases, so that 'M has the same cardinality as N' is to be regarded simply as another way of saying 'M and N are equinumerous'. It seems not to be until 1883 that he treats cardinals as distinct objects to be manipulated in their own right. The definition of cardinal which Cantor eventually gave in 1895 is characteristic of his attitude to foundational questions: '"Power" or "cardinal number" of M is the name we give to the general concept which emerges from the set M with the help of our active capacity for thought by abstracting from the nature of its distinct elements and from the order in which they are given.'

Frege criticized Cantor's psychological approach to definitions in *Die Grundlagen der Arithmetik* (1884) and proposed to define

$$\text{card}(A) = \{X : X \text{ is equinumerous with } A\}. \tag{5.2.1}$$

This definition does not work in our theory since if the object defined in (5.2.1) exists and $A \neq \varnothing$ then the mapping $\mathfrak{P}(\text{card}(A)) \to \text{card}(A)$ given by $X \mapsto A \times \{X\}$ is injective and therefore has a surjective right inverse (cf. exercise 1 of §2.2) contrary to Cantor's theorem. The version of Frege's idea (due to Tarski (1955) and Scott (1955)) which we have adopted here depends vitally on the well-founded structure of the hierarchy of days. In Zermelo's (1908*a*) system, which does not have such a structure, one has to postulate an extra primitive term card(A) with the property that

$$\text{card}(A) = \text{card}(B) \Leftrightarrow A \text{ and } B \text{ are equinumerous.} \tag{5.2.2}$$

No such term is definable in Zermelo's system (Lévy 1969).

In stronger systems, however, other methods of defining cardinality are possible. In a system such as Bourbaki's (1954) with the STRONG IOTA-CONSISTENCY AXIOM one can (using an idea due essentially to Dedekind (1888)) let

$$\text{card}(A) = \iota X(X \text{ and } A \text{ are equinumerous})$$

and obtain in addition to (5.2.2) the neat but unnecessary property that

$$\text{card}(\text{card}(A)) = \text{card}(A). \tag{5.2.3}$$

In a system with the axiom of choice and the AXIOM OF REPLACEMENT, on the other hand, one can show that every set is equinumerous with at least one pseudo-ordinal and then let card(A) be the least such pseudo-ordinal, thereby obtaining properties (5.2.2) and (5.2.3) again.

In the remainder of this chapter the lower-case Fraktur letters $\mathfrak{a}$, $\mathfrak{b}$, $\mathfrak{c}$, *etc. will denote cardinals.*

Remark 1 Every cardinal $\mathfrak{a}$ is non-empty and if $A \in \mathfrak{a}$ then $\text{card}(A) = \mathfrak{a}$. Moreover, if $X \in \text{card}(A)$ then $\text{D}(X) \subseteq \text{D}(A)$ and so $\text{card}(A) \subseteq \mathfrak{P}(\text{D}(A))$.

Definition

Suppose that $\mathfrak{a} = \text{card}(A)$ and $\mathfrak{b} = \text{card}(B)$. We write $\mathfrak{a} \leq \mathfrak{b}$ if there exists an injective mapping $A \to B$.

Justification It is clear that this definition does not depend on the choice of the representative collections A and B. For if $\text{card}(A) = \text{card}(A')$ and $\text{card}(B) = \text{card}(B')$ then there are bijections $f : A \to A'$ and $g : B \to B'$ [proposition 5.2.1]. So if $i : A \to B$ is an injection then $g \circ i \circ f^{-1} : A' \to B'$ is one too; and if $i' : A' \to B'$ is an injection then $g^{-1} \circ i' \circ f : A \to B$ is one too. □

Exercises

1. Show that $\mathfrak{a} \leq \mathfrak{b} \Rightarrow \text{D}(\mathfrak{a}) \subseteq \text{D}(\mathfrak{b})$.
2. Show that $\{\mathfrak{a} : \mathfrak{a} \text{ is a cardinal}\}$ does not exist.

Lemma 5.2.2

(*a*) $\mathfrak{a} \leq \mathfrak{a}$.
(*b*) $\mathfrak{a} \leq \mathfrak{b}$ *and* $\mathfrak{b} \leq \mathfrak{c} \Rightarrow \mathfrak{a} \leq \mathfrak{c}$.
(*c*) $\mathfrak{a} \leq \mathfrak{b}$ *and* $\mathfrak{b} \leq \mathfrak{a} \Rightarrow \mathfrak{a} = \mathfrak{b}$.

Proof (a) and (b) are trivial; (c) is a rewording of Bernstein's equinumerosity theorem. □

Lemma 5.2.3

$\mathfrak{a}$ is small (i.e. a set) iff there exists a set A such that $\mathfrak{a} = \text{card}(A)$.

Necessity Let $\mathfrak{a}$ be a small cardinal. Now there exists a collection $A \in \mathfrak{a}$. It is clear that $\text{card}(A) = \mathfrak{a}$ and that A is a set.

Sufficiency This follows from the observation we made in remark 1 that the day on which $\text{card}(A)$ is born is no later than the day after the birthday of A. □

Definition

The class of all small cardinals is denoted **Cn**.

Proposition 5.2.4

(**Cn**, $\leq$) *is a partially ordered class.*

Proof Immediate [lemma 5.2.2]. □

Remark 2 We have not claimed that $(\mathbf{Cn}, \leq)$ is a totally ordered class, i.e. that for any $\mathfrak{a}, \mathfrak{b} \in \mathbf{Cn}$ either $\mathfrak{a} \leq \mathfrak{b}$ or $\mathfrak{b} \leq \mathfrak{a}$: this claim cannot be proved from the axioms now at our disposal; in fact it is equivalent to an extra set-theoretic assumption called the 'axiom of choice' (see §7.3).

Proposition 5.2.5

$\mathbf{Cn}$ *does not have a maximal element.*

Proof It follows from Cantor's theorem that $\text{card}(A) < \text{card}(\mathfrak{P}(A))$ for every set A. □

Corollary 5.2.6

If $\mathfrak{a} \in \mathbf{Cn}$ *then* $\mathfrak{a} < \text{card}(\mathbf{D})$.

Proof If $\mathfrak{a} \in \mathbf{Cn}$ then there exists a set A such that $\text{card}(A) = \mathfrak{a}$ [lemma 5.2.3], and so $\mathfrak{a} \leq \text{card}(\mathbf{D})$ (since $A \subseteq \mathbf{D}$). But if $\text{card}(\mathbf{D}) = \mathfrak{a} \in \mathbf{Cn}$, then $\text{card}(\mathbf{D})$ is the greatest element of $\mathbf{Cn}$, contrary to proposition 5.2.5. □

Proposition 5.2.7

A class $B \subseteq \mathbf{Cn}$ *is a set iff it has an upper bound in* $\mathbf{Cn}$.

Necessity Suppose that B is a set and let $\mathfrak{c} = \text{card}(\mathrm{D}(B))$. Then $\mathfrak{c} \in \mathbf{Cn}$ [lemma 5.2.3]; moreover, if $\mathfrak{a} \in B$ then for any $A \in \mathfrak{a}$ we have $A \subseteq \mathrm{D}(B)$, so that

$$\mathfrak{a} = \text{card}(A) \leq \text{card}(\mathrm{D}(B)) = \mathfrak{c}.$$

Thus $\mathfrak{c}$ is an upper bound for B.

Sufficiency Suppose $\mathfrak{c} \in \mathbf{Cn}$ is an upper bound for B and choose a set C such that $\text{card}(C) = \mathfrak{c}$ [lemma 5.2.3]. Then

$$B \subseteq \{\text{card}(X) : X \in \mathfrak{P}(C)\},$$

and so B is a set. □

Corollary 5.2.8

$\mathbf{Cn}$ *is a proper class.*

Proof Since $\mathbf{Cn}$ does not have a maximal element [proposition 5.2.5], it in particular has no upper bound and is therefore not a set [proposition 5.2.7]. But all the elements of $\mathbf{Cn}$ are sets and so $\mathbf{Cn}$ is certainly a class. □

5.3 Finite and infinite

We have already defined what it means for a collection to be 'finite' or 'countable'. Let us now investigate how these concepts fit into the theory of cardinals.

Definition

We let $\aleph_0 = \text{card}(\omega)$; for each $n \in \omega$ we let $|n| = \text{card}(\boldsymbol{n})$. We say that a collection A *has n elements* if $\text{card}(A) = |n|$, i.e. if A and $\boldsymbol{n}$ are equinumerous. We say that A is *infinite* if there exists an injective mapping $A \to A$ which is not bijective.

The collection of all n-element subcollections of A is denoted $\mathfrak{F}_n(A)$.

By a slight abuse of language we say that $\mathfrak{a} = \text{card}(A)$ is *finite* [respectively *infinite, countable*] if A is finite [respectively infinite, countable]. This definition is evidently independent of the choice of the representative collection A.

Remark 1 It is clear that a collection A is finite iff it has n elements for some $n \in \omega$. Hence $\mathfrak{F}(A) = \bigcup_{n \in \omega} \mathfrak{F}_n(A)$.

Remark 2 Many authors use the word 'infinite' to mean 'not finite'; such authors typically use 'Dedekind-infinite' for what we have called 'infinite' and 'Dedekind-finite' for 'not infinite'.

Example 1 $\mathfrak{F}_0(A) = \{\varnothing\}$, $\mathfrak{F}_1(A) = \{\{a\} : a \in A\}$.

Example 2 ω is infinite since the mapping $s : \omega \to \omega$ is injective but not bijective.

Example 3 A cardinal $\mathfrak{a}$ is finite iff $\mathfrak{a} = |n|$ for some $n \in \omega$.

Proposition 5.3.1

A collection A is countable iff $\text{card}(A) \leq \aleph_0$.

Necessity Suppose that A is countable. If A is empty, then there is trivially an injective mapping $A \to \omega$. If not, then there is a surjective mapping $f : \omega \to A$. For each $x \in A$ the set $f^{-1}[x]$ is therefore non-empty and has a least element $g(x)$. The mapping $g : A \to \omega$ thus defined is evidently injective. Hence $\text{card}(A) \leq \aleph_0$.

Sufficiency Suppose that $\text{card}(A) \leq \aleph_0$. So there is an injective mapping $g : A \to \omega$. If A is empty, then it is certainly countable. If not, then we can choose an element $a \in A$: now if $n \in g[A]$ let $f(n)$ be the unique $x \in A$ such that $g(x) = n$; and if $n \in \boldsymbol{\omega} - g[A]$ let $f(n) = a$. The mapping $f : \boldsymbol{\omega} \to A$ thus defined is evidently surjective and so A is countable in this case too. □

Theorem 5.3.2

A collection A is finite iff $\operatorname{card}(A) < \aleph_0$.

Necessity Suppose that A is finite. So A has n elements for some $n \in \boldsymbol{\omega}$. Now $\boldsymbol{n} \subseteq \boldsymbol{\omega}$, so that obviously $\operatorname{card}(A) \leq \aleph_0$. But if $\operatorname{card}(A) = \aleph_0$, then $\boldsymbol{\omega}$ is finite and therefore has a maximal (hence greatest) element [lemma 4.4.1], which is absurd. So $\operatorname{card}(A) < \aleph_0$ as required.

Sufficiency Suppose that $\operatorname{card}(A) < \aleph_0$ but A is not finite. Note first that A is non-empty and countable [proposition 5.3.1] and so there exists a sequence (x_n) whose image is A. It is clear that if we can define an injective mapping $g : \boldsymbol{\omega} \to A$, we shall have the contradiction we require (since then $\operatorname{card}(A) \geq \aleph_0$). We do this by general recursion. If $g(m)$ has been defined for $m < n$, then $A - g[\boldsymbol{n}]$ is not empty since A is not finite, and so we can let $g(n)$ be the element x_r of $A - g[\boldsymbol{n}]$ of least possible index r. It is clear that the mapping $g : \boldsymbol{\omega} \to A$ thus defined is injective. □

Theorem 5.3.3

A collection A is infinite iff $\operatorname{card}(A) \geq \aleph_0$.

Necessity Suppose that A is infinite. So there exists a mapping $f : A \to A$ which is injective but not bijective, and therefore there is an element $a \in A - f[A]$. Now $\operatorname{Cl}_f(a)$ is a subalgebra of (A, f) which is isomorphic to $(\boldsymbol{\omega}, s)$ [corollary 4.1.5] and so

$$\operatorname{card}(A) \geq \operatorname{card}(\operatorname{Cl}_f(a)) = \operatorname{card}(\boldsymbol{\omega}) = \aleph_0.$$

Sufficiency Suppose that $\operatorname{card}(A) \leq \aleph_0$. So there is an injective mapping $g : A \to \boldsymbol{\omega}$. If A is empty, then it is certainly countable. If not, then we can

$$f(x) = \begin{cases} g(g^{-1}(x) + 1) & \text{if } x \in g[\boldsymbol{\omega}] \\ x & \text{if } x \in A - g[\boldsymbol{\omega}] \end{cases}$$

is clearly injective, but it is not bijective since $g(0) \in A - f[A]$. □

Remark 3 One obvious consequence of these two theorems is that a collection cannot be both finite and infinite: the question whether it can be neither will be (partially) answered in §5.5.

Proposition 5.3.4

$m \leq n \Leftrightarrow |m| \leq |n|$.

Proof If $m \leq n$ then $\boldsymbol{m} \subseteq \boldsymbol{n}$ and so $|m| \leq |n|$. If, on the other hand, $|m| \leq |n|$ but $m > n$, then there exists an injective mapping $\boldsymbol{m} \to \boldsymbol{n}$: this mapping restricts to an injective but not bijective mapping $\boldsymbol{n} \to \boldsymbol{n}$. Hence $\boldsymbol{n}$ is both finite and infinite, contradicting theorems 5.3.2 and 5.3.3. □

Corollary 5.3.5

$|m| = |n| \Rightarrow m = n.$

Proof Immediate [proposition 5.3.4]. □

It is a trivial consequence of the definitions that a finite collection has n elements for *at least* one $n \in \boldsymbol{\omega}$: what this corollary tells us is that this is true for *exactly* one $n \in \boldsymbol{\omega}$; we can therefore unambiguously refer to this n as the *number of elements* in the collection.

Proposition 5.3.6

The number of elements in a collection A is the least $n \in \boldsymbol{\omega}$ such that $\mathfrak{F}_{n+1}(A) = \varnothing$.

Proof Exercise. □

Exercises

1. Show that these three assertions are equivalent:
   (i) $\text{card}(A) = \aleph_0$;
   (ii) A is countable and infinite;
   (iii) A is countable and not finite.
   A is said to be *countably infinite*—'denumerable' according to some authors—if it satisfies these conditions.
2. Show that $\mathfrak{a}$ is not finite iff $\mathfrak{a} > |n|$ for all $n \in \boldsymbol{\omega}$.
3. Let (A, f) be a successor algebra.
   (a) If A is finite, show that $f : A \to A$ is injective iff it is surjective.
   (b) If there exists an an element $a \in f[A]$ such that $A = \text{Cl}_f(a)$, show that $f : A \to A$ is bijective. [First use the simple recursion principle to show that A is finite.]
4. Prove proposition 5.3.6.

5.4 Cardinal arithmetic

Definition

Suppose that $\mathfrak{a}, \mathfrak{b} \in \mathbf{Cn}$. So there exist sets A and B such that $\mathfrak{a} = \text{card}(A)$ and $\mathfrak{b} = \text{card}(B)$. We let

$$\mathfrak{a} + \mathfrak{b} = \text{card}(A \cup B)$$
$$\mathfrak{a}\mathfrak{b} = \text{card}(A \times B)$$
$$\mathfrak{b}^{\mathfrak{a}} = \text{card}({}^{A}B).$$

Justification It is easy to show that these definitions do not depend on our choice of the representative sets A and B. □

Proposition 5.4.1

Suppose $m, n \in \boldsymbol{\omega}$:

$$|m+n| = |m| + |n|;$$

$$|mn| = |m||n|;$$

$$|m^n| = |m|^{|n|}.$$

Proof Immediate [proposition 4.4.2(c)–(e)]. □

We have already shown [proposition 5.3.4] that the mapping from $\boldsymbol{\omega}$ onto the set of all finite cardinals given by $n \mapsto |n|$ is an order isomorphism. Proposition 5.4.1 shows that it also preserves the arithmetic structure of the natural numbers. So it seems reasonable to suppose that little confusion will arise if we denote the cardinal $|n|$ by the symbol n; we shall adopt this policy from now on.

Proposition 5.4.2

Suppose $\mathfrak{a}, \mathfrak{b}, \mathfrak{c} \in \mathbf{Cn}$:

(*a*) $(\mathfrak{a} + \mathfrak{b}) + \mathfrak{c} = \mathfrak{a} + (\mathfrak{b} + \mathfrak{c})$;

(*b*) $\mathfrak{a} + \mathfrak{b} = \mathfrak{b} + \mathfrak{a}$;

(*c*) $\mathfrak{a} + 0 = \mathfrak{a}$;

(*d*) $\mathfrak{a} \geq \mathfrak{b} \Leftrightarrow (\exists \mathfrak{d} \in \mathbf{Cn})(\mathfrak{a} = \mathfrak{b} + \mathfrak{d})$;

(*e*) $\mathfrak{b} \leq \mathfrak{c} \Rightarrow \mathfrak{a} + \mathfrak{b} \leq \mathfrak{a} + \mathfrak{c}$;

(*f*) $(\mathfrak{a}\mathfrak{b})\mathfrak{c} = \mathfrak{a}(\mathfrak{b}\mathfrak{c})$;

(*g*) $\mathfrak{a}\mathfrak{b} = \mathfrak{b}\mathfrak{a}$;

(*h*) $\mathfrak{a}0 = 0$; $\mathfrak{a}1 = \mathfrak{a}$; $\mathfrak{a}2 = \mathfrak{a} + \mathfrak{a}$;

(*i*) $\mathfrak{a}(\mathfrak{b} + \mathfrak{c}) = \mathfrak{a}\mathfrak{b} + \mathfrak{a}\mathfrak{c}$;

(*j*) $\mathfrak{b} \leq \mathfrak{c} \Rightarrow \mathfrak{a}\mathfrak{b} \leq \mathfrak{a}\mathfrak{c}$;

(*k*) $(\mathfrak{a}^{\mathfrak{b}})^{\mathfrak{c}} = \mathfrak{a}^{\mathfrak{b}\mathfrak{c}}$;

(*l*) $(\mathfrak{a}\mathfrak{b})^{\mathfrak{c}} = \mathfrak{a}^{\mathfrak{c}}\mathfrak{b}^{\mathfrak{c}}$;

(*m*) $\mathfrak{a}^{\mathfrak{b}+\mathfrak{c}} = \mathfrak{a}^{\mathfrak{b}}\mathfrak{a}^{\mathfrak{c}}$;

(*n*) $\mathfrak{a}^0 = 1$; $\mathfrak{a}^1 = \mathfrak{a}$; $\mathfrak{a}^2 = \mathfrak{a}\mathfrak{a}$;

(*o*) *if* $\mathfrak{a} \leq \mathfrak{c}$ *and* $\mathfrak{b} \leq \mathfrak{d}$, *then* $\mathfrak{a}^{\mathfrak{c}} \leq \mathfrak{b}^{\mathfrak{d}}$;

(*p*) $\mathfrak{a} < 2^{\mathfrak{a}}$;

(*q*) $\mathfrak{a} \geq \aleph_0$ *iff* $\mathfrak{a} = \mathfrak{a} + 1$.

Proof Part (p) follows from proposition 5.1.3(a) and theorem 5.1.4; part (q) follows from theorem 5.3.3. All the other parts are straightforward to prove. As an illustration let us prove part (k). To do this it will be enough to show that if A, B, C are sets then ${}^C({}^BA)$ is equinumerous with ${}^{C\times B}A$: this is achieved by observing that if we take each function $f \in {}^C({}^BA)$ to the function in ${}^{C\times B}A$ given by $(c,b) \mapsto f(c)(b)$ then we obtain a one-to-one correspondence; its inverse takes each function $g \in {}^{C\times B}A$ to the function in ${}^C({}^BA)$ given by $c \mapsto (b \mapsto g(c,b))$. □

Proposition 5.4.3

(*a*) $2\aleph_0 = \aleph_0$.
(*b*) $\aleph_0{}^2 = \aleph_0$.

Proof Immediate [theorem 4.4.2(a) and (b)]. □

There are a great many cardinal identities which follow more or less straightforwardly from the properties we have listed.

Example 1 To prove that $\aleph_0{}^{\aleph_0} = 2^{\aleph_0}$ observe that

$$\begin{aligned}2^{\aleph_0} &\le \aleph_0{}^{\aleph_0} \text{ [proposition 5.4.2(o)]}\\ &\le (2^{\aleph_0})^{\aleph_0} \text{ [proposition 5.4.2(o) and (p)]}\\ &= 2^{\aleph_0{}^2} \text{ [proposition 5.4.2(k) and (n)]}\\ &= 2^{\aleph_0} \text{ [proposition 5.4.3(b)].}\end{aligned}$$

Example 2 The identity $\aleph_0 2^{\aleph_0} = 2^{\aleph_0}$ follows from the fact that

$$\begin{aligned}2^{\aleph_0} &\le \aleph_0 2^{\aleph_0} \text{ [proposition 5.4.2(j)]}\\ &\le \aleph_0{}^{\aleph_0} 2^{\aleph_0} \text{ [proposition 5.4.2(j) and (o)]}\\ &= (2\aleph_0)^{\aleph_0} \text{ [proposition 5.4.2(l)]}\\ &= \aleph_0{}^{\aleph_0} \text{ [proposition 5.4.3(a)]}\\ &= 2^{\aleph_0} \text{ [example 1].}\end{aligned}$$

Example 3 $\mathfrak{a}$ is infinite iff $\mathfrak{a} + \aleph_0 = \mathfrak{a}$. For if $\mathfrak{a}$ is infinite then $\mathfrak{a} \ge \aleph_0$ [theorem 5.3.3] and so there exists $\mathfrak{b} \in \mathbf{Cn}$ such that $\mathfrak{a} = \mathfrak{b} + \aleph_0$ [proposition 5.4.2(d)], whence

$$\begin{aligned}\mathfrak{a} + \aleph_0 &= (\mathfrak{b} + \aleph_0) + \aleph_0\\ &= \mathfrak{b} + (\aleph_0 + \aleph_0) \text{ [proposition 5.4.2(a)]}\\ &= \mathfrak{b} + 2\aleph_0 \text{ [proposition 5.4.2(h)]}\\ &= \mathfrak{b} + \aleph_0 \text{ [proposition 5.4.3(a)]}\\ &= \mathfrak{a}.\end{aligned}$$

If conversely $\mathfrak{a} + \aleph_0 = \mathfrak{a}$, then $\mathfrak{a} \ge \aleph_0$ [proposition 5.4.2(d)] and so $\mathfrak{a}$ is infinite [theorem 5.3.3].

Exercise 1. Suppose that $\mathfrak{a}, \mathfrak{b} \in \mathbf{Cn}$.
(a) $\mathfrak{a} \geq 2^{\aleph_0}$ iff $\mathfrak{a} + 2^{\aleph_0} = \mathfrak{a}$;
(b) if $\mathfrak{a} \geq 2^{\aleph_0} \geq \mathfrak{b}$, then $\mathfrak{a} + \mathfrak{b} = \mathfrak{a}$;
(c) $\mathfrak{a} = 2^{\aleph_0}$ iff $\mathfrak{a} \geq \aleph_0$ and $\mathfrak{a} + \aleph_0 = 2^{\aleph_0}$;
(d) if $2 \leq \mathfrak{a} \leq \mathfrak{b} = \mathfrak{b}^2$, then $\mathfrak{a}^{\mathfrak{b}} = 2^{\mathfrak{b}}$.

Proposition 5.4.4

$\text{card}(\mathbf{R}) = 2^{\aleph_0}$.

Proof First notice that the mapping $f : \mathbf{R} \to \mathfrak{P}(\mathbf{Q})$ given by $f(a) = \{x \in \mathbf{Q} : x < a\}$ is injective since $\mathbf{Q}$ is dense in $\mathbf{R}$, and so

$$\text{card}(\mathbf{R}) \leq \text{card}(\mathfrak{P}(\mathbf{Q})) = 2^{\text{card}(\mathbf{Q})} = 2^{\aleph_0}. \tag{5.4.1}$$

Now choose three elements $a_0, b_0, c_0 \in \mathbf{Q}$ such that $a_0 < b_0 < c_0$. Suppose that $s = (s_n)_{n \in \omega}$ is an element of ${}^{\omega}\mathbf{2}$ and define sequences $(a_n(s))_{n \in \omega}$, $(b_n(s))_{n \in \omega}$, and $(c_n(s))_{n \in \omega}$ in $\mathbf{Q}$ recursively as follows. First let $a_0(s) = a_0$, $b_0(s) = b_0$, and $c_0(s) = c_0$. Choose $b_{n+1}(s)$ such that

$$a_n(s) < b_{n+1}(s) < c_n(s):$$

and if $s_n = 0$ then let $a_{n+1}(s) = a_n(s)$ and choose $c_{n+1}(s)$ such that

$$a_{n+1}(s) < c_{n+1}(s) < b_{n+1}(s);$$

but if $s_n = 1$ then let $c_{n+1}(s) = c_n(s)$ and choose $a_{n+1}(s)$ such that

$$b_{n+1}(s) < a_{n+1}(s) < c_{n+1}(s).$$

(In each case make the choice according to some definite rule determined in advance: for example, fix beforehand a sequence (x_n) whose image is $\mathbf{Q}$ and specify that each choice of an element x_r of $\mathbf{Q}$ should be made so that the index r is as small as possible.) Now $(a_n(s))$ is an increasing sequence whose range is bounded. So $g(s) = \sup_{n \in \omega} a_n(s)$ exists. We shall show now that the mapping $g : {}^{\omega}\mathbf{2} \to \mathbf{R}$ thus defined is injective. Suppose that s, s' are distinct elements of ${}^{\omega}\mathbf{2}$. Let n_0 be the least index such that $s_{n_0} \neq s_{n_0}'$ and suppose for the sake of argument that $s_{n_0} = 0$ and $s_{n_0}' = 1$. Then for all $n \geq n_0 + 1$ it is the case that

$$a_n(s) < c_n(s) \leq c_{n_0+1}(s) < b_{n_0+1}(s),$$

whereas

$$a_{n_0+1}(s') > b_{n_0+1}(s') = b_{n_0+1}(s).$$

Therefore

$$g(s) \leq c_{n_0+1}(s) < b_{n_0+1}(s) < a_{n_0+1}(s') \leq g(s').$$

This shows that $g : {}^{\omega}\mathbf{2} \to \mathbf{R}$ is injective. Consequently,

$$\text{card}(\mathbf{R}) \geq \text{card}({}^{\omega}\mathbf{2}) = 2^{\aleph_0}. \tag{5.4.2}$$

It now follows from (5.4.1) and (5.4.2) that card(**R**) = $2^{\aleph_0}$ [Bernstein's equinumerosity theorem]. □

Remark 1 In this proof we have paid the price for omitting to mention field structure in our specification of **R**. If we could use this structure then the function $(s_n) \mapsto \Sigma_{n \in \omega} s_n 3^{-n-1}$ would provide us with a suitable injective mapping $^{\omega}\mathbf{2} \to \mathbf{R}$ much more directly. (The image of this function is called 'Cantor's ternary set'.)

Remark 2 Because of proposition 5.4.4 collections of cardinality $2^{\aleph_0}$ are sometimes said to have the *power of the continuum* ('power' here being an old synonym for 'cardinality').

Corollary 5.4.5

R *is not countable.*

Proof Immediate [propositions 5.4.4 and 5.4.2(p)]. □

The results on cardinal arithmetic which we have established in this section make it possible to calculate the cardinalities of a wide variety of collections.

Example 4 $\mathrm{card}(\mathfrak{P}(\omega)) = \mathrm{card}(^{\omega}\mathbf{2}) = 2^{\aleph_0}$.

Example 5 Let us determine the cardinality of the set $\mathscr{A}$ of all equivalence relations on ω. First notice that every equivalence relation on ω is a subset of $\omega \times \omega$, and so

$$\mathrm{card}(\mathscr{A}) \leq \mathrm{card}(\mathfrak{P}(\omega \times \omega)) = 2^{\aleph_0{}^2} = 2^{\aleph_0}. \tag{5.4.3}$$

Now consider the mapping $f : \mathfrak{P}(\omega) \to \mathscr{A}$ which takes a set $B \subseteq \omega$ to the equivalence relation on ω whose equivalence classes are B and the singletons $\{n\}$ for $n \in \omega - B$: it is not injective since it does not distinguish between 1-element subsets of ω; but the restriction $f|\mathfrak{P}(\omega) - \mathfrak{F}_1(\omega) \to \mathscr{A}$ is injective, and so

$$\mathrm{card}(\mathscr{A}) \geq \mathrm{card}(\mathfrak{P}(\omega) - \mathfrak{F}_1(\omega)).$$

Now $\mathrm{card}(\mathfrak{F}_1(\omega)) = \aleph_0$, $\mathrm{card}(\mathfrak{P}(\omega)) = 2^{\aleph_0}$ (see example 4) and $\mathrm{card}(\mathfrak{P}(\omega) - \mathfrak{F}_1(\omega)) \geq \aleph_0$. Hence by exercise 1(c) $\mathrm{card}(\mathfrak{P}(\omega) - \mathfrak{F}_1(\omega)) = 2^{\aleph_0}$. Thus

$$\mathrm{card}(\mathscr{A}) \geq 2^{\aleph_0}. \tag{5.4.4}$$

It follows from (5.4.3) and (5.4.4) that $\mathrm{card}(\mathscr{A}) = 2^{\aleph_0}$ [Bernstein's equinumerosity theorem].

Example 6 $\mathrm{card}(\mathbf{R}^n) = (2^{\aleph_0})^n = 2^{n\aleph_0} = 2^{\aleph_0}$ if $n \geq 1$.

Remark 3 It follows from example 6 that there exists a one-to-one correspondence between **R** and $\mathbf{R}^n$ for each $n \geq 1$. This result astonished Cantor when he first discovered it: *Je le vois, mais je ne le crois pas*, he wrote to Dedekind (Cantor 1877).

Equinumerosity The observation that the set of natural numbers is equinumerous with a proper subset of itself is usually attributed to Galileo but it can also be found in the fourteenth-century writings of Albert of Saxony (see Maier 1949, p. 170f.). The conclusion Galileo drew from this observation was that 'the attributes of equal, greater, and less have no place in infinite quantities, but only in bounded ones' (Galileo 1638, p. 79). Bolzano in his book *Paradoxien des Unendlichen*, published posthumously in 1851, repeated Galileo's observation and noted that this property is characteristic of infinite sets. He also defined the property of equinumerosity, gave a rather general definition of the notion of 'set', and presented a non-mathematical proof that there exist infinitely many sets. Moreover, he undoubtedly had an influence on Cantor, who mentioned *Paradoxien des Unendlichen* in glowing terms, and on Dedekind, who in *Was sind und was sollen die Zahlen*? (1888) adopted Bolzano's characterization of infinite collections as a definition and reproduced his 'proof' that there exist infinitely many sets. However, those who look for the outlines of Cantor's theory of cardinals in Bolzano's work will be largely disappointed: he had no more awareness than Galileo that equinumerosity can be used to measure the size of infinite sets and without this insight to guide him he based what superficially appears to be an attempt at cardinal arithmetic on nothing more than a crude analogy with the finite case.

In any case, Cantor did not at first approach the concept of equinumerosity from this abstract point of view. He showed how the rationals can be enumerated as a sequence in 1873 and then asked, in a letter to Dedekind (1873*a*), whether the reals can also be enumerated. His proof that they cannot, which he discovered a week or so later (letter to Dedekind, 1873*b*) and published in 1874, gave birth to the study of cardinal arithmetic, although it was not until 1883 and 1895 that he published the general theory which we have presented in this chapter.

Exercises

2. Find the cardinalities of the following sets:
   (a) the set of functions from $\mathbf{R}$ to $\mathbf{R}$;
   (b) the set of open subsets of $\mathbf{R}$ with its usual topology*;
   (c) the set of continuous functions from $\mathbf{R}$ to $\mathbf{R}$;
   (d) the set of subsets of $\boldsymbol{\omega}$ with more than one element;
   (e) the set $\mathfrak{F}(\boldsymbol{\omega})$ of finite subsets of $\boldsymbol{\omega}$;
   (f) the set of infinite subsets of $\boldsymbol{\omega}$;
   (g) the set of bijective mappings $\boldsymbol{\omega} \to \boldsymbol{\omega}$.

* We assume here an elementary knowledge of the topology of the real line.

3. Show that the collection $B = \{\mathfrak{b} \in \mathbf{Cn} : \mathfrak{b}^2 = \mathfrak{b}\}$ is a proper class. [*Hint* $2^{\mathfrak{a}\aleph_0} \in B$ for all $\mathfrak{a} \in \mathbf{Cn}$.]
4. Suppose that $\mathfrak{a} \in \mathbf{Cn}$.
   (a) Prove that if $\mathfrak{a}$ is not finite then $2^{2^{\mathfrak{a}}}$ is infinite. [If $\mathrm{card}(A) = \mathfrak{a}$, consider the function $n \mapsto \mathfrak{F}_n(A)$.]
   (b) Prove that if $\mathfrak{a}$ is not finite then $2^{2^{\mathfrak{a}}} \geq 2^{\aleph_0}$.

5.5 The axiom of countable choice

Is every set either finite or infinite? Is a countable union of countable sets necessarily countable? The axioms we have stated so far are not sufficient to settle these questions. The best we can do is to show that affirmative answers to them follow from an extra set-theoretic assumption called the axiom of countable choice.

Definition

The *axiom of countable choice* asserts that for every small sequence (A_n) of non-empty sets there exists a sequence (x_n) such that $x_n \in A_n$ for all $n \in \omega$.

Remark 1 This axiom cannot be proved in **GA** (Fraenkel 1922*b*) or even in **G** (Cohen 1963–4). Most mathematicians regard its truth as self-evident; some do not. In deference to those in the latter category (and because of the intrinsic interest in singling out the results which depend on it) we shall not regard countable choice as an axiom of basic set theory but explicitly mention it each time that it is used.

Remark 2 The axiom of countable choice is used quite frequently in analysis. Typically one has under consideration a formula $\Phi(n, x)$ expressing a property (dependent on n) of the real number x. Provided that one has the theorem

$$(\forall n \in \omega)(\exists x \in \mathbf{R})\Phi(n, x)$$

the sets $A_n = \{x \in \mathbf{R} : \Phi(n, x)\}$ are non-empty and therefore the axiom of countable choice ensures the existence of a sequence (x_n) in $\mathbf{R}$ such that $\Phi(n, x)$ for all $n \in \omega$.

In analysis textbooks uses of the axiom are rarely signalled explicitly. The one we have just described might be introduced as follows: 'For each natural number n let us choose a real number x_n such that $\Phi(n, x_n)$.'

Remark 3 It is sometimes possible to avoid using the axiom by giving an explicit rule for defining the elements of the sequence in question. For example, if we have the theorem

$$(\forall n \in \omega)(\exists x \in \mathbf{Q})\Phi(n, x)$$

then we can define explicitly a sequence (x_n) in $\mathbf{Q}$ such that $\Phi(n, x_n)$ for all $n \in \omega$: to do this we note that $\mathbf{Q}$ is countable (unlike $\mathbf{R}$) and so there exists a sequence (a_r) whose image is $\mathbf{Q}$; if we let r_n be the least element of the set $\{r \in \omega : \Phi(n, a_r)\}$ (which is non-empty by hypothesis) and let $x_n = a_{r_n}$ for each $n \in \omega$, then we obtain a sequence (x_n) with the property we require.

Theorem 5.5.1

It follows from the axiom of countable choice that every set is either finite or infinite.

Proof Suppose that A is a set which is not finite. Then $\mathfrak{F}_n(A) \neq \varnothing$ for all $n \in \omega$ [proposition 5.3.6]. So there is a sequence $(A_n)_{n \in \omega}$ such that $A_n \in \mathfrak{F}_n(A)$ for each $n \in \omega$ [axiom of countable choice]. Now

$$\operatorname{card}\left(\bigcup_{r<n} A_{2^r}\right) \leq \sum_{r<n} 2^r = 2^n - 1 < 2^n = \operatorname{card}(A_{2^n}),$$

so that

$$A_{2^n} - \bigcup_{r<n} A_{2^r} \neq \varnothing.$$

Hence [axiom of countable choice again] there is a mapping $g : \omega \to A$ such that

$$g(n) \in A_{2^n} - \bigcup_{r<n} A_{2^r}$$

for each $n \in \omega$. This mapping is evidently injective and therefore A is infinite [theorem 5.3.3]. □

To put it another way, the axiom of countable choice implies that every small cardinal is comparable with $\aleph_0$.

Theorem 5.5.2

It follows from the axiom of countable choice that if (A_n) is a small sequence of countable sets then $\bigcup_{n \in \omega} A_n$ is also a countable set.

Proof By hypothesis the set of one-to-one functions from A_n to ω is non-empty for each $n \in \omega$; so [axiom of countable choice] there is a sequence (f_n) such that f_n is a one-to-one function from A_n to ω for each $n \in \omega$. Now the mapping $f : \biguplus_{n \in \omega} A_n \to \omega \times \omega$ defined by $f(x, n) = (f_n(x), n)$ is clearly injective. If we compose this with a bijective mapping $\omega \times \omega \to \omega$ [proposition 4.4.2(a)], we obtain an injective mapping $\biguplus_{n \in \omega} A_n \to \omega$ and deduce that $\biguplus_{n \in \omega} A_n$ is countable. But there is a canonical surjective mapping dom: $\biguplus_{n \in \omega} A_n \to \bigcup_{n \in \omega} A_n$ and therefore $\bigcup_{n \in \omega} A_n$ is also countable. □

The axiom of countable choice The first implicit use of the axiom of countable choice seems to have been in a proof, which appeared in Heine (1872) with an ascription to Cantor, that a real-valued function which is sequentially continuous at a point is also continuous at it. Between then and the end of the century it was used implicitly on many occasions by Cantor, Dedekind, Borel, Baire, and others. Only Peano and his colleagues in Turin seem to have commented explicitly on its use: in 1890 Peano wrote, 'One may not apply an infinite number of times an arbitrary law according to which a class is made to correspond to an individual of that class'; and in 1896 Bettazzi criticized Dedekind's (1888) proof that every set is either finite or infinite because 'one must choose an object (correspondence) arbitrarily in each of the *infinite* sets, which does not seem rigorous; unless one wishes to accept as a postulate that such a choice can be carried out—something, however, which seems ill-advised to us.'

Nevertheless, the axiom of countable choice has a certain plausibility, since it appears to be only medically, rather than logically, impossible to make an infinite number of choices in a finite time by the device of performing each choice in half the time it took to perform the previous one (cf. Russell 1936, pp. 143–4). Moreover, anyone who rejects the axiom must accept that classical analysis is considerably harmed as a result: not only may there be subsets of **R** which are neither finite nor infinite (Cohen 1966, p. 138) but, rather more strikingly, the continuum may be a countable union of countable sets (Feferman and Levy 1963).

Exercises

1. Assuming the axiom of countable choice, prove that for every small cardinal $\mathfrak{a} > \aleph_0$ there exists exactly one $\mathfrak{b} \in \mathbf{Cn}$ such that $\mathfrak{a} = \aleph_0 + \mathfrak{b}$.
2. Prove (without using the axiom of countable choice) that **R** is not a countable union of finite sets.

6 Ordinals

The simple and general principles of induction are powerful tools for proving things about the natural numbers: we intend now to investigate ways in which they can be generalized to apply to a very much wider class of ordered collections than the subsets of $\boldsymbol{\omega}$. The basis of our study will be the observation that the general principle of induction follows trivially from the least element principle and can therefore be applied to any ordered collection which shares this property: we call such collections 'well ordered'. Our strategy will be to apply the same techniques to the study of isomorphism between well ordered sets as we used in Chapter 4 to investigate equinumerosity between sets: just as the work of Chapter 4 produced an arithmetic of cardinals, in this chapter we shall get an arithmetic of *ordinals*.

6.1 Isomorphism properties of well ordered collections

Definition

If $(A, \leq)$ is a partially ordered collection and $a \in A$, we let

$$\operatorname{seg}_A(a) = \{x \in A : x < a\}.$$

Lemma 6.1.1

If $(A, \leq)$ is a partially ordered collection then these three assertions are equivalent:

(*i*) *every non-empty subcollection of A has a minimal element*;

(*ii*) *every subcollection of A which has a strict upper bound in A has a strict supremum in A*;

(*iii*) *if $B \subseteq A$ and $(\forall a \in A)(\operatorname{seg}_A(a) \subseteq B \Rightarrow a \in B)$, then $B = A$.*

Proof Exercise. □

Definition

If $(A, \leq)$ is a [partially] ordered collection which satisfies the equivalent conditions of lemma 6.1.1 above, then we say that $\leq$ is a [*partial*] *well-ordering* on A and that $(A, \leq)$ is a [*partially*] *well ordered collection.* The collection A is said to be *well-orderable* if there exists a well-ordering on A

Remark 1 Any partially ordered collection in which every non-empty subcollection has a *least* element is totally (and hence well) ordered.

Remark 2 Obviously every subcollection of a [partially] well ordered collection is [partially] well ordered by the inherited ordering. In particular, the initial subcollection $\mathrm{seg}_A(a)$ is [partially] well ordered if A is.

Remark 3 A partially well-ordered collection cannot contain (the image of) a strictly decreasing sequence. (See §7.2 for a discussion of the converse.)

Remark 4 Rather more generally, a relation r on A is said to be *well founded* if for every $B \subseteq A$ we have

$$B \neq \varnothing \Rightarrow B - r[B] \neq \varnothing.$$

It is easy to see that a [partial] order relation $\leq$ on A is a [partial] well-ordering iff the associated strict [partial] order $<$ is well founded.

Remark 5 If A and B are equinumerous then A is well-orderable iff B is. So whether or not a collection is well-orderable depends only on its cardinality.

Remark 6 $(A, \leq)$ is said to be *Noetherian* if A^{op} is partially well ordered, i.e. if every non-empty subcollection of A has a maximal element.

Suppose that $(A, \leq)$ is a well ordered collection. A has a least element $\perp$ iff A is non-empty. A need not have a greatest element: if $a \in A$ then either a is the greatest element of A or there exist elements of A greater than a, in which case the least of these is the unique successor of a, denoted a^+. An element of $A - \{\perp\}$ need not be the successor of any other element of A: if it is not, it is called a *limit point* of A; this is the case iff $a = \sup \mathrm{seg}_A(a)$.

Example 1 The motivating example of a well ordered set is $(\boldsymbol{\omega}, \leq)$: it has no limit points and no greatest element.

Example 2 Every finite [partially] ordered collection is [partially] well ordered [theorem 4.4.1]: indeed an ordered collection is finite iff both it and its opposite are well ordered (exercise 7 of §4.4).

Example 3 The proper class $\mathscr{D}$ of all small days is well ordered by inclusion: its limit points are the small limit days.

Theorem 6.1.2

If $(A, \leq)$ is a well ordered collection and $f : A \to A$ is a strictly increasing mapping, then $x \leq f(x)$ for all $x \in A$.

Proof Suppose not. So there exist elements $x \in A$ such that $x > f(x)$; let x_0 b the least such x. Then

$$f(x_0) \le f(f(x_0)) < f(x_0).$$

Contradiction. □

Corollary 6.1.3

If $(A, \le)$ is a well ordered collection and $B \subseteq A$, then these three assertion are equivalent:
(i) B is a proper initial subcollection of A;
(ii) there exists an element $a \in A$ such that $B = \mathrm{seg}_A(a)$;
(iii) B is an initial subcollection of A which is not isomorphic to A.

(i) ⇒ (ii) If B is a proper initial subcollection of A then $A - B$ is non-empt and therefore has a least element a; plainly $B = \mathrm{seg}_A(a)$.
(ii) ⇒ (iii) Suppose that $B = \mathrm{seg}_A(a)$. If there is an isomorphism $f: A \to B$ then $f(a) \ge a$ [theorem 6.1.2] and so $f(a) \notin B$, which is absurd.
(iii) ⇒ (i) Obvious. □

Corollary 6.1.4

Let $(A, \le)$ and $(B, \le)$ be well ordered collections. If there is an isomorphisn from $(A, \le)$ to $(B, \le)$ then it is unique.

Proof Suppose that $f, g: A \to B$ are isomorphisms. Then $g \circ f^{-1}: B \to B$ i strictly increasing: so for all $x \in A$ we have

$$f(x) \le g(f^{-1}(f(x))) = g(x) \quad \text{[theorem 6.1.2]}$$

and similarly $g(x) \le f(x)$. Hence $f = g$. □

Theorem 6.1.5

If $(A, \le)$ and $(B, \le)$ are well ordered collections, then either A is isomorphi to an initial subcollection of B or B is isomorphic to an initial subcollectio of A (or both).

Proof Let

$$f = \{(x, y) \in A \times B: \mathrm{seg}_A(x) \text{ and } \mathrm{seg}_B(y) \text{ are isomorphic}\}.$$

Note first that if $(x_1, y_1), (x_2, y_2) \in f$ then $x_1 < x_2 \Leftrightarrow y_1 < y_2$ since if, fo example, $x_1 < x_2$ and $y_1 \ge y_2$ then $\mathrm{seg}_B(y_1)$ is isomorphic to a prope initial subcollection of itself, contrary to corollary 6.1.3. It follows fron this that f is the graph of a strictly increasing mapping and that dom[f and im[f] are initial subcollections of A and B respectively. Suppos now, if possible, that dom[f] $\ne A$ and im[f] $\ne B$. Then let a be th least element of A − dom[f] and b be the least element of B − im[f]

Evidently, $\mathrm{dom}[f] = \mathrm{seg}_A(a)$ and $\mathrm{im}[f] = \mathrm{seg}_B(b)$, and so f is the graph of an isomorphism between $\mathrm{seg}_A(a)$ and $\mathrm{seg}_B(b)$. Hence $b = f(a)$. Contradiction. Thus f is the graph of the isomorphism we require. □

Proposition 6.1.6

If $(A, \leq)$ is well ordered and $(B, \leq)$ is partially ordered, then a mapping $f: A \to B$ is normal [respectively strictly normal] iff it satisfies:

(1) *if $a \in A$ is a limit point of A then $f(a) = \sup_{x<a} f(x)$;*

(2) *$f(a^+) \geq f(a)$ [respectively $f(a^+) > f(a)$] for all a in A apart from the greatest element of A (if it exists).*

Necessity Obvious.

Sufficiency Let us prove first that f is increasing. For if it is not, then there exists $a \in A$ such that for some $b < a$ we have $f(b) \not\leq f(a)$: choose the element a as small as possible. There are two cases to consider:

(I) a is a limit point of A. In this case

$$f(a) = \sup_{x<a} f(x) \not\leq f(a).$$

Contradiction.

(II) $a = c^+$ for some $c \in A$. In this case for all $x < a$ we have $x \leq c$ and hence $f(x) \leq f(c)$ (by the minimality of a), so that $c = b$ and hence $f(c) \not\leq f(a)$. But

$$f(a) = f(c^+) \geq f(c).$$

Contradiction.

Therefore f is indeed increasing. [The proof that f is strictly increasing under the stronger version of the second hypothesis is similar.]

Now let us show that f is normal. Suppose that C is a non-empty bounded subcollection of A and let $c = \sup C$. Now if $x \in C$ then $x \leq c$ and so $f(x) \leq f(c)$; hence $f(c)$ is an upper bound for $f[C]$. Now consider two possibilities:

(I) $c \in C$. In this case $f(c) \in f[C]$ and so $f(c)$ is the supremum of $f[C]$.

(II) $c \notin C$. In this case c is a limit point of A. Now if $x < c$ then there exists $y \in C$ such that $x \leq y$ and so if z is an upper bound for $f[C]$ in B then $f(x) \leq f(y) \leq z$. Hence

$$f(c) = \sup_{x<c} f(x) \leq z.$$

It follows that $f(c)$ is the supremum of $f[C]$ in this case too.

Thus f is [strictly] normal. □

Proposition 6.1.7

If $(A, \leq)$ is a well ordered collection and $f : A \to A$ is strictly normal, then for every $a \geq f(\perp)$ there exists a greatest element $b \in A$ such that $f(b) \leq a$.

Proof If a is the greatest element of A, then it is clearly the element we are looking for. If not, then $f(a^+) > f(a) \geq a$ and so there exist elements $c \in A$ such that $f(c) > a$: choose the least such element c. Now $c > \perp$ since $f(c) > f(\perp)$. And if c were a limit point of A, then we would have

$$f(c) = \sup_{x<c} f(x) \leq a < f(c),$$

which is absurd. So there exists $b \in A$ such that $c = b^+$. Clearly $f(b) \leq a$ and if $x > b$ then $x \geq c$ so that $f(x) \geq f(c) > a$. Thus b is the element we want. □

Proposition 6.1.8

If the partially ordered collection $(A, \leq)$ is the union of a collection $\mathscr{B}$ of initial subcollections each of which is partially well ordered by the inherited partial order, then A is also partially well ordered.

Proof Suppose that A has a non-empty subcollection C with no minimal element and choose an element $x \in C$. Now $x \in B$ for some $B \in \mathscr{B}$. So $B \cap C$ is non-empty and therefore has a minimal element y since B is partially well ordered. Since C has no minimal element, there exists $z \in C$ such that $z < y$; moreover $z \in B$ since B is an initial subcollection of A. This contradicts the minimality of y in $B \cap C$. Hence $(A, \leq)$ is partially well ordered. □

Exercises

1. Prove lemma 6.1.1.
2. Suppose that $(A, \leq)$ is a partially ordered collection.
   (a) Show that $(A, \leq)$ is partially well ordered iff $\mathrm{seg}_A(a)$ is partially well ordered for every $a \in A$.
   (b) Show that A is finite iff $\mathfrak{P}(A)$ is partially well ordered by inclusion. [*Sufficiency* Consider $\{B \in \mathfrak{P}(A) : A - B \text{ is finite}\}$.]
3. In the notation of exercise 6 of §2.6 show that $\leq$ is a partial well-ordering on D.
4. Let $(A, \leq)$ be a partially ordered collection.
   (a) Show that these two assertions are equivalent:
      (i) $\mathrm{seg}_A(a)$ is well ordered for every $a \in A$;
      (ii) $(A, \leq)$ is partially well ordered and every directed subcollection of A is totally ordered by the inherited ordering.
   $(A, \leq)$ is said to be a *tree* if it satisfies these conditions.
   (b) If A is a tree show that B is a maximal totally ordered subcollection of A iff it is a totally ordered initial subcollection of A with no strict upper bound in A. B is said to be a *branch* of A if it satisfies these conditions.
5. (a) Prove that every countable directed collection has a well ordered cofinal subcollection.

(b) Show that $\mathfrak{F}(\mathbf{R})$ is directed by inclusion but does not contain a cofinal chain.

6. Show that every subset of **R** which is well ordered by the inherited ordering is countable. [Associate a rational with each element of the subset.]
7. A totally ordered collection $(A, \leq)$ is said to be *perfectly ordered* if it satisfies:
   A has a least element;
   every element of A except the greatest (if there is one) has a unique successor;
   every element of A can be obtained by applying the successor operation a finite number of times to either the least element of A or a limit point of A.
   Show that every well ordered collection is perfectly ordered but that the converse does not hold.
8. Show that every well founded relation is a strict subordering.

.2 Definition of ordinals

efinition

If (A, r) is a structure, then the collection $\langle (B, s) : (B, s)$ is isomorphic to $(A, r)\rangle$ is denoted $\operatorname{ord}(A, r)$ and called the *type* of (A, r).

roposition 6.2.1

If (A, r) *and* (B, s) *are structures then* (A, r) *is isomorphic to* (B, s) *iff* $\operatorname{ord}(A, r) = \operatorname{ord}(B, s)$.

roof Straightforward. □

efinition

The type of a well ordered collection is called an *ordinal.*

In the rest of this chapter the Greek letters α, β, γ, *etc. will denote ordinals.*

efinition

If $\alpha = \operatorname{ord}(A, \leq)$ and $\beta = \operatorname{ord}(B, \leq)$, we shall write $\alpha \leq \beta$ if there exists an isomorphism of $(A, \leq)$ onto an initial subcollection of $(B, \leq)$.

Justification It is easy to see that this definition does not depend on our choice of the representatives $(A, \leq)$ and $(B, \leq)$. □

roposition 6.2.2

(a) $\alpha \leq \beta$ *and* $\beta \leq \gamma \Rightarrow \alpha \leq \gamma$
(b) $\alpha \leq \alpha$
(c) $\alpha \leq \beta$ *and* $\beta \leq \alpha \Rightarrow \alpha = \beta$
(d) $\alpha \leq \beta$ *or* $\beta \leq \alpha$.

roof Parts (a) and (b) are trivial; (c) follows from corollary 6.1.3 and (d) from theorem 6.1.5. □

Proposition 6.2.3

If the partially ordered collection $(A, \leq)$ *is the union of a chain* $\mathscr{B}$ *of initia subcollections each of which is well ordered by the inherited partial order then* A *is also well ordered and*

$$\operatorname{ord}(A, \leq) = \sup_{B \in \mathscr{B}} \operatorname{ord}(B, \leq_B).$$

Proof Straightforward [proposition 6.1.8]. □

Definition

$\boldsymbol{\alpha} = \{\beta : \beta < \alpha\}.$

Justification To see that this exists we need only note that

$$\{\beta : \beta < \alpha\} = \{\beta \in \mathfrak{P}(D(\alpha)) : \beta < \alpha\}. \quad \square$$

Remark 1 This definition may seem somewhat eccentric to anyone who is use to the von Neumann definition of the ordinals—what we have calle 'pseudo-ordinals'—since with that definition $\boldsymbol{\alpha} = \alpha$. Of course, for us th two concepts are different (although closely related, as the next theorem shows).

Theorem 6.2.4

$(\boldsymbol{\alpha}, \leq)$ *is a well ordered collection and* $\operatorname{ord}(\boldsymbol{\alpha}, \leq) = \alpha$.

Proof If $(A, \leq)$ is a well ordered collection such that $\operatorname{ord}(A, \leq) = \alpha$, then it i easy to check that the mapping $f : A \to \boldsymbol{\alpha}$ defined by $f(x) = \operatorname{ord}(\operatorname{seg}_A(x), \leq$ is an isomorphism: it is strictly increasing since if $x < y$ then $\operatorname{seg}_A(x)$ is a initial subcollection of, but not isomorphic to, $\operatorname{seg}_A(y)$ [corollary 6.1.3 and so $f(x) < f(y)$; it is surjective since if $\beta < \alpha$ then there exists $x \in$ such that

$$\beta = \operatorname{ord}(\operatorname{seg}_A(x), \leq) = f(x) \quad \text{[corollary 6.1.3].} \quad \square$$

Theorem 6.2.5 (*Burali-Forti's paradox*)

$\{\alpha : \alpha$ *is an ordinal*$\}$ *does not exist.*

Proof Suppose on the contrary that $A = \{\alpha : \alpha \text{ is an ordinal}\}$ exists. The $A = \bigcup_{\alpha \in A} \boldsymbol{\alpha}$ and each $\boldsymbol{\alpha}$ is well ordered [theorem 6.2.4]; hence A i well ordered [proposition 6.2.3]. So if $\alpha = \operatorname{ord}(A, \leq)$, then $\operatorname{ord}(\boldsymbol{\alpha}, \leq) =$ [theorem 6.2.4] and therefore $(\boldsymbol{\alpha}, \leq)$ is isomorphic to $(A, \leq)$ [propositio 6.2.1]. But $\boldsymbol{\alpha}$ is a proper initial subcollection of A (since $\alpha \in A$) and i therefore not isomorphic to it [corollary 6.1.3]. Contradiction. □

Burali-Forti's paradox If we call the order types of perfectly ordered collections 'quasi-ordinals' then what Burali-Forti (1897*a*) proved was that the collection of all quasi-ordinals is not totally ordered in its natural partial order. He also asserted that every quasi-ordinal is an ordinal, thereby contradicting Cantor's (1897) theorem that the ordinals are totally ordered. But there is a mistake here: Burali-Forti had misunderstood Cantor's definition of a well ordering and it is in fact the converse implication—every ordinal is a quasi-ordinal—which holds (cf. exercise 7 of §6.1). So the contradiction dissolves: there is nothing absurd about a non-totally ordered collection having a totally ordered subcollection. Burali-Forti (1897*a*) corrected his error and added the rather enigmatic remark that 'the reader can check which propositions in [Burali-Forti 1897*a*] are verified also by the well ordered classes'. However, he did not take his own advice: it was left to Russell to resurrect the paradox in 1903 by noticing that Burali-Forti's argument shows without any essential change that the *ordinals* are not totally ordered. Because of Russell's attribution, this has become universally known as Burali-Forti's paradox, although Burali-Forti himself denied as late as 1906 that any contradiction was involved (letter to Couturat, 1906)*. Cantor had in any event discovered the paradox independently by 1899 (letters to Dedekind, 1899*a*, *b*).

Definition

$|\alpha| = \text{card}(\boldsymbol{\alpha})$.

Lemma 6.2.6

These three assertions are equivalent:

(i) α *is small;*

(ii) $|\alpha|$ *is small;*

(iii) $\boldsymbol{\alpha}$ *is a set.*

(i) $\Rightarrow$ (iii) This follows from the fact that $\boldsymbol{\alpha} \subseteq \mathfrak{P}(D(\alpha))$.

(iii) $\Rightarrow$ (ii) Immediate [lemma 5.2.3].

(ii) $\Rightarrow$ (i) Suppose that $|\alpha|$ is small. So there exists a set A such that $\text{card}(A) = |\alpha| = \text{card}(\boldsymbol{\alpha})$ [lemma 5.2.3]. Since A and $\boldsymbol{\alpha}$ are equinumerous, there exists a well-ordering $\leq$ on A such that

$$\text{ord}(A, \leq) = \text{ord}(\boldsymbol{\alpha}, \leq) = \alpha \quad \text{[theorem 6.2.4]}.$$

Hence α is small. □

Definition

The least ordinal which is not small is denoted On.

* The letter as it appears is hopelessly confused and may have been mistranslated.

Justification If every ordinal were small, then $\{\alpha : \alpha \text{ is an ordinal}\}$ woul be a subclass of **D**, contradicting Burali-Forti's paradox. Hence there exis ordinals which are not small. In particular, it is easy to see that there exist a unique least ordinal which is not small. □

Proposition 6.2.7

On *is the proper class of all small ordinals.*

Proof The fact that **On** is a proper class follows from lemma 6.2.6; the rest i obvious. □

Proposition 6.2.8

A class $B \subseteq \mathbf{On}$ *is a set iff it is bounded in* **On**.

Necessity Suppose that $B \subseteq \mathbf{On}$ is a set. So $D(B)$ is a set and therefor there exists a small ordinal α not in $D(B)$ (since otherwise **On** would be set). Now if $\beta \in B$ and $\beta \not\leq \alpha$ then

$$\alpha \subseteq D(\alpha) \subseteq D(\beta) \in D(B)$$

and hence $\alpha \in D(B)$, which contradicts our choice of α. So α is an uppe bound for B in **On** as required.

Sufficiency If B is bounded in **On**, then it has a small strict supremum (say), and so $B \subseteq \boldsymbol{\beta}$: but $\boldsymbol{\beta}$ is a set [lemma 6.2.6] and hence B is a set. □

The least ordinal is denoted 0. There is evidently no greatest ordinal. Fo any ordinal α we let α^+ denote the least ordinal $> \alpha$ (i.e. the successor o α). A non-zero ordinal which is not the successor of any other ordinal i called a *limit ordinal.*

We say that α is *finite* or *infinite* according as $|\alpha|$ is finite or infinite. W let $\omega_0 = \text{ord}(\boldsymbol{\omega}, \leq)$, so that $|\omega_0| = \aleph_0$ and therefore ω_0 is the least infinit ordinal. If for each $n \in \boldsymbol{\omega}$ we let $n_0 = \text{ord}(\boldsymbol{n}, \leq)$, then the mapping $\boldsymbol{\omega} \to \boldsymbol{\omega}$ given by $n \mapsto n_0$ is an isomorphism: from now on we shall identify th natural number n with the finite ordinal n_0 and write ω for ω_0.

Remark 2 A family $(x_\alpha)_{\alpha<\beta}$ indexed by $\boldsymbol{\beta}$ for some ordinal β is called a *transfinit sequence.* It should be clear that a collection is well-orderable iff ther exists a transfinite sequence whose range it is.

Remark 3 The element of **On** form a transfinite sequence: it starts like the natur numbers but carries on beyond them. Here is a fragment of it. (Th notation involved—addition, multiplication, etc.—will be defined late in this chapter.)

$$0, 1, 2, 3, \ldots, \omega, \omega + 1, \omega + 2, \ldots, \omega 2, \omega 2 + 1, \ldots, \omega 3, \ldots,$$
$$\omega^{(2)}, \omega^{(2)} + 1, \ldots, \omega^{(2)} + \omega, \omega^{(2)} + \omega + 1, \ldots, \omega^{(2)} + \omega 2,$$
$$\omega^{(2)} + \omega 2 + 1, \ldots, \omega^{(2)} + \omega 3, \ldots, \omega^{(2)}2, \omega^{(2)}2 + 1, \ldots,$$
$$\omega^{(3)}, \ldots, \omega^{(\omega)}, \ldots, \omega^{(\omega^{(\omega)})}, \ldots, \varepsilon_0, \ldots, \varepsilon_1, \ldots, \varepsilon_\omega, \ldots,$$
$$\omega_1, \ldots, \omega_2, \ldots, \mathrm{On}, \ldots.$$

Lemma 6.2.9

If $\alpha \neq 0$ then these three assertions are equivalent:
(i) α *is a limit ordinal*;
(ii) $(\forall \beta)(\beta < \alpha \Rightarrow \beta^+ < \alpha)$;
(iii) $\alpha = \sup_{\beta < \alpha} \beta$.

(i) ⇒ (ii) Suppose $\beta < \alpha$. Then certainly $\beta^+ \leq \alpha$. But if $\beta^+ = \alpha$, then α is not a limit. So $\beta^+ < \alpha$ as required.
(ii) ⇒ (iii) let $\gamma = \sup_{\beta < \alpha} \beta$. Then $\gamma \leq \alpha$. But if $\gamma < \alpha$, then by hypothesis $\gamma^+ < \alpha$ and so $\gamma^+ \leq \gamma$, which is absurd. So $\gamma = \alpha$.
(iii) ⇒ (i) Suppose that α is not a limit. Then there exists an ordinal γ such that $\alpha = \gamma^+$. Hence

$$\gamma = \sup_{\beta < \alpha} \beta = \alpha = \gamma^+,$$

which is absurd. □

Example 1 The least limit ordinal is ω.

Example 2 On is a limit ordinal.

Example 3 If On_0 denotes the ordinal of the class $\mathscr{D}$ of all small days ordered by inclusion, then there is exactly one isomorphism of $\mathbf{On}_0$ onto $\mathscr{D}$: the graph of this isomorphism is denoted $\alpha \mapsto \mathrm{D}_\alpha$ and D_α is called the *αth day*; the inverse graph is denoted $D \mapsto \rho(D)$ and $\rho(D)$ is called the *date* of the day D. It is sometimes convenient to extend ρ so that it is defined on the whole of $\mathbf{D}$: we can do this by letting $\rho(A) = \rho(\mathrm{D}(A))$ for each $A \in \mathbf{D}$; in other words, $\rho(A)$ is the date of A's birthday. (More prosaic authors call it the 'rank' of A.) It is clear that On_0 is a limit ordinal and $\omega < \mathrm{On}_0 \leq \mathrm{On}$. But the axioms of basic set theory do not allow us to say any more about the value of On_0: it is consistent with them that On_0 is the next limit ordinal after ω, whereas On is certainly much larger than that. However, if we assume REPLACEMENT, then $\mathrm{On}_0 = \mathrm{On}$.

Pseudo-ordinals The first complete treatment of ordinal arithmetic using pseudo-ordinals was given by von Neumann in 1923. The notion of pseudo-ordinal had already been identified by Zermelo in some unpublished work about 1915 (see Hallett 1984, pp. 270–80) and again by

Mirimanoff (1917), but von Neumann was the first to see how REPLACE-MENT legitimizes the method by ensuring that $\mathrm{On}_0 = \mathrm{On}$. (Fraenke who had stated the AXIOM OF REPLACEMENT in 1922 before vo Neumann, later recorded his surprise at von Neumann's discovery of th link with pseudo-ordinals (Fraenkel 1967, p. 169).)

We now consider the generalizations to arbitrary ordinals of the simpl and general principles of induction and recursion which we proved i Chapter 5.

Proposition 6.2.10

If $A \subseteq \boldsymbol{\alpha}$ and if $\boldsymbol{\beta} \subseteq A \Rightarrow \beta \in A$, then $A = \boldsymbol{\alpha}$.

Proof Immediate [lemma 6.1.1 and theorem 6.2.4]. □

Corollary 6.2.11 (*General principle of transfinite induction*)

If $\Phi(\alpha)$ is a formula such that

$$(\forall\beta)((\forall\gamma < \beta)\Phi(\gamma) \Rightarrow \Phi(\beta)),$$

then $(\forall\alpha)\Phi(\alpha)$.

Proof Let $A = \{\beta < \alpha : \Phi(\beta)\}$; then $A = \boldsymbol{\alpha}$ [proposition 6.2.10] and so $\Phi(\alpha)$ holds. □

Proposition 6.2.12

Suppose that $A \subseteq \boldsymbol{\alpha}$ has these properties:

$0 \in A$;

$\beta \in A \Rightarrow \beta^+ \in A$ *for every $\beta < \alpha$;*

$\boldsymbol{\lambda} \subseteq A \Rightarrow \lambda \in A$ *for every limit ordinal $\lambda < \alpha$.*

Then $A = \boldsymbol{\alpha}$.

Proof Suppose, if possible, that $A \subset \boldsymbol{\alpha}$. So $\boldsymbol{\alpha} - A$ is non-empty and therefore ha a least element [theorem 6.2.4]. The first property shows that this elemen is not zero, the second that it is not a successor, and the third that it i not a limit ordinal. Contradiction. □

Corollary 6.2.13 (*Simple principle of transfinite induction*)

Suppose that $\Phi(\alpha)$ is a formula such that:

$\Phi(0)$;

$(\forall\beta)(\Phi(\beta) \Rightarrow \Phi(\beta^+))$;

if λ is a limit ordinal, then $(\forall\beta < \lambda)\Phi(\beta) \Rightarrow \Phi(\lambda)$.

Then $(\forall\alpha)\Phi(\alpha)$.

roof This follows from proposition 6.2.12 in the same way as corollary 6.2.11 follows from proposition 6.2.10. □

heorem 6.2.14 (*General principle of transfinite recursion*)
If for each $\beta < \alpha$ there is a mapping $s_\beta : \mathfrak{P}(A) \to A$, then there exists exactly one mapping $g : \boldsymbol{\alpha} \to A$ such that $g(\beta) = s_\beta(g[\boldsymbol{\beta}])$ for all $\beta < \alpha$.

Uniqueness This follows straightforwardly from the general principle of transfinite induction.

Existence We shall prove the existence of g by simple transfinite induction on α. This is trivial if $\alpha = 0$. Then if there is a mapping $g : \boldsymbol{\beta} \to A$ with the required property, we can extend g to $\boldsymbol{\beta}^+$ by letting $g(\beta) = s_\beta(g[\boldsymbol{\beta}])$. Finally, suppose that λ is a limit ordinal and that for each $\beta < \lambda$ there exists a (necessarily unique) mapping $g_\beta : \boldsymbol{\beta} \to A$ such that $g_\beta(\gamma) = s_\gamma(g_\beta[\boldsymbol{\gamma}])$ for all $\gamma < \beta$. Any two of these functions g_β must agree on the intersection of their domains. So if $g = \bigcup_{\beta<\lambda} g_\beta$ then [lemma 2.2.1] g is a function and

$$\mathrm{dom}[g] = \bigcup_{\beta<\lambda} \mathrm{dom}[g_\beta] = \{\gamma : (\exists \beta < \lambda)(\gamma < \beta)\} = \boldsymbol{\lambda}.$$

It is easy to see that $g : \boldsymbol{\lambda} \to A$ is the mapping we require. This completes the proof by simple transfinite induction. □

roposition 6.2.15 (*Simple principle of transfinite recursion*)
If (A, f) is a successor algebra, $a \in A$, and $s : \mathfrak{P}(A) \to A$ is a mapping, then for each ordinal α there exists a unique mapping $g : \boldsymbol{\alpha} \to A$ such that:

$$g(0) = a;$$
$$g(\beta^+) = f(g(\beta)) \quad \text{if } \beta^+ < \alpha;$$
$$g(\lambda) = s(g[\boldsymbol{\lambda}]) \quad \text{for every limit ordinal } \lambda < \alpha.$$

roof The proof is similar to that of the general principle of transfinite recursion. □

emark 4 In the case when $\alpha = \omega$ all these results reduce to the corresponding principles proved in §§4.1 and 4.3; when we use them, we shall sometimes exploit the same sort of temporal terminology as we mentioned in Chapter 4.

xample 4 As an illustration of the use of simple transfinite recursion, let us show that if $A \times A$ is well-orderable then every partial ordering $\leq$ on A is contained in a total ordering $\leq'$ on A. First we choose a transfinite sequence $((x_\alpha, y_\alpha))_{\alpha<\beta}$ whose range is $A \times A$. (This is possible since $A \times A$ is well-orderable.) We then define recursively a family $(\leq_\alpha)_{\alpha\leq\beta}$ of partial orderings on A containing $\leq$ as follows. First let $\leq_0 = \leq$. Now suppose that $\leq_\alpha$ has been defined and $\alpha < \beta$: if $\leq_\alpha \cup \{(x_\alpha, y_\alpha)\}$ is contained in a partial ordering on A, let $\leq_{\alpha^+}$ be the smallest such partial ordering (i.e. the partial ordering generated by $\leq_\alpha \cup \{(x_\alpha, y_\alpha)\}$); if there is no such partial ordering, let $\leq_{\alpha^+} = \leq_\alpha$. Finally, if λ is a limit ordinal let $\leq_\lambda = \bigcup_{\alpha<\lambda} \leq_\alpha$. (The existence of the family $(\leq_\alpha)_{\alpha\leq\beta}$ thus described is a consequence of the simple principle of transfinite recursion.) If we now let $\leq' = \leq_\beta$ it is clear that $\leq'$ contains $\leq$. Moreover, it is straightforward to check that $\leq'$ is a maximal element of the collection $\mathrm{Po}(A)$ of all partial orderings on A and is therefore a total ordering on A by exercise 4(b) of §2.6.

Exercises

1. (a) If $(A, \leq)$ is a partially ordered collection, show that these three assertions are equivalent:
   (i) $(A, \leq)$ is partially well ordered;
   (ii) there exist exactly one ordinal α and surjective mapping $\rho_A : A \to \alpha$ (called the *rank mapping* of A) such that $\rho_A(a) = \sup^+_{x<a} \rho_A(x)$ for all $a \in A$;
   (iii) there exists a strictly increasing mapping from A to a well-ordered collection.
   (b) If $(A, \leq)$ is a tree, show that $\rho_A(a) = \mathrm{ord}(\mathrm{seg}_A(a), \leq)$ for all $a \in A$.
2. If $f : \mathbf{On} \to \mathbf{On}$ is normal, $\alpha \in \mathbf{On}$, and $\{f^n(\alpha) : n \in \boldsymbol{\omega}\}$ is bounded in $\mathbf{On}$, show that $\sup_{n\in\boldsymbol{\omega}} f^n(\alpha)$ is the least fixed point of f which is $\geq \alpha$.
3. Show that the birthday of α is no more than 4 days after the birthday of $|\alpha|$.

6.3 Ordinal addition

In the next three sections we shall define operations of addition, multiplication, and exponentiation for ordinals. We shall take as our model the recursive definitions of the corresponding operations for natural numbers (§4.2): the form the extended definitions should take at successor ordinals is clear; the form of the additional clause defining the behaviour of the operations at limit ordinals is also clear if we require that the operations must all be normal in their second variable. In the case of ordinal addition, for example, we want a mapping $\mathbf{On} \times \mathbf{On} \to \mathbf{On}$ given by $(\alpha, \beta) \mapsto \alpha + \beta$ such that:

$$\alpha + 0 = \alpha;$$

$$\alpha + \beta^+ = (\alpha + \beta)^+;$$

$$\alpha + \lambda = \sup_{\beta<\lambda} (\alpha + \beta) \quad \text{if } \lambda \text{ is a limit ordinal.}$$

The easiest way to show that such a function exists is to proceed synthetically in much the same way as we defined the cardinal operations in §5.3: it will then be a straightforward matter to observe that the resulting function does indeed have the recursive properties we require.

)efinition

If $(A, \leq)$ and $(B, \leq)$ are totally ordered collections, then we define the *ordered sum* $A + B$ of A with B to be the ordered pair $(A \cup B, \leq)$ where $A \cup B = (A \times \{0\}) \cup (B \times \{1\})$ as usual and $(x, i) \leq (y, j)$ iff $i = 0$ and $j = 1$, **or** $i = j = 0$ and $x \leq y$, **or** $i = j = 1$ and $x \leq y$.

This definition amounts to placing a copy of B *after* a copy of A; the ordinary sum of the structures $(A, \leq)$ and $(B, \leq)$ (defined in Chapter 2) amounts to placing a copy of B *beside* a copy of A. It is clear that the ordered sum of two totally ordered collections is totally ordered (unlike the ordinary sum). However, in order for it to be useful to us here it must also be *well ordered* if its components are. Fortunately, this is easily seen to be the case.

.emma 6.3.1

The ordered sum of two well ordered collections is well ordered.

'roof Suppose that C is a non-empty subcollection of $A \cup B$. So

$$C = D \cup E = (D \times \{0\}) \cup (E \times \{1\})$$

for some $D \subseteq A$ and $E \subseteq B$. If D is non-empty then it has a least element a in A, in which case $(a, 0)$ is the least element of C; if D is empty then E is non-empty and has a least element b in B, in which case $(b, 1)$ is the least element of C. □

)efinition

If $\alpha = \text{ord}(A, \leq)$ and $\beta = \text{ord}(B, \leq)$ then we define $\alpha + \beta$ to be the ordinal of the ordered sum of $(A, \leq)$ and $(B, \leq)$.

Justification This definition is justified by lemma 6.3.1 and the fact that it evidently does not depend on our choice of $(A, \leq)$ and $(B, \leq)$. □

'roposition 6.3.2

$|\alpha + \beta| = |\alpha| + |\beta|$ for all $\alpha, \beta \in \mathbf{On}$.

'roof Trivial. □

'roposition 6.3.3

The mapping $\mathbf{On} \times \mathbf{On} \to \mathbf{On}$ *given by* $(\alpha, \beta) \mapsto \alpha + \beta$ *satisfies*:

(*a*) $\alpha + 0 = \alpha$;

(*b*) $\alpha + \beta^+ = (\alpha + \beta)^+$;

(*c*) $\alpha + \lambda = \sup_{\beta < \lambda}(\alpha + \beta)$ *if* λ *is a limit ordinal.*

'roof Straightforward. □

Remark 1 It is clear from proposition 6.3.3 that for finite ordinals (i.e. natura numbers) this definition of addition coincides with the one given in §4.

Remark 2 Notice that $\alpha + 1 = \alpha + 0^+ = (\alpha + 0)^+ = \alpha^+$. Consequently propositio 6.3.3(b) can be rewritten as

$$\alpha + (\beta + 1) = (\alpha + \beta) + 1.$$

Corollary 6.3.4

For each $\alpha \in \mathbf{On}$ the mapping $\mathbf{On} \to \mathbf{On}$ given by $\beta \mapsto \alpha + \beta$ is strictl normal.

Proof Immediate [propositions 6.1.6 and 6.3.3]. □

Theorem 6.3.5 (*Subtraction algorithm*)

If $\alpha, \beta \in \mathbf{On}$ and $\beta \leq \alpha$ then there exists a unique $\rho \in \mathbf{On}$ such that $\alpha = \beta + \rho$

Existence There exists a greatest ordinal ρ such that $\beta + \rho \leq \alpha$ [corollar 6.3.4 and proposition 6.1.7]. But if $\beta + \rho < \alpha$, then

$$\beta + (\rho + 1) = (\beta + \rho) + 1 \leq \alpha,$$

contradicting our choice of ρ. Hence $\beta + \rho = \alpha$ as required.

Uniqueness This follows from the injectivity of the mapping $\rho \mapsto \beta +$ [corollary 6.3.4]. □

Exercise 1. Show that if $\rho < \tau$ in the above decomposition then $\alpha < \beta + \tau$.

Remark 3 If $\beta \leq \alpha$ then we write $\alpha - \beta$ to denote the unique ordinal ρ such tha $\alpha = \beta + \rho$.

Proposition 6.3.6

Suppose that $\alpha, \beta, \gamma \in \mathbf{On}$:

(*a*) $\beta < \gamma \Rightarrow \alpha + \beta < \alpha + \gamma$;

(*b*) $\alpha + \beta = \alpha + \gamma \Rightarrow \beta \Rightarrow \gamma$;

(*c*) *if $B \subseteq \mathbf{On}$ is non-empty and bounded then* $\alpha + \sup_{\beta \in B} \beta = \sup_{\beta \in B} (\alpha + \beta$

(*d*) $\alpha + (\beta + \gamma) = (\alpha + \beta) + \gamma$;

(*e*) $\alpha \leq \beta \Rightarrow \alpha + \gamma \leq \beta + \gamma$;

(*f*) $\alpha \leq \beta \Leftrightarrow (\exists \delta \in \mathbf{On})(\beta = \alpha + \delta)$;

(*g*) $\alpha < \beta \Leftrightarrow (\exists \delta \in \mathbf{On} - \{0\})(\beta = \alpha + \delta)$.

Proof (a), (b), and (c) Immediate [corollary 6.3.4].

(d) We use transfinite induction on γ. The case $\gamma = 0$ is trivial. If $\alpha +$ $(\beta + \gamma) = (\alpha + \beta) + \gamma$, then

$$\begin{aligned}\alpha + (\beta + (\gamma + 1)) &= \alpha + ((\beta + \gamma) + 1) \\ &= (\alpha + (\beta + \gamma)) + 1 \\ &= ((\alpha + \beta) + \gamma) + 1 \\ &= (\alpha + \beta) + (\gamma + 1).\end{aligned}$$

And if λ is a limit ordinal such that $\alpha + (\beta + \gamma) = (\alpha + \beta) + \gamma$ for all $\gamma < \lambda$, then

$$\begin{aligned}\alpha + (\beta + \lambda) &= \sup_{\gamma<\lambda} (\alpha + (\beta + \gamma)) \quad [\text{(c)}] \\ &= \sup_{\gamma<\lambda} ((\alpha + \beta) + \gamma) \\ &= (\alpha + \beta) + \lambda \quad [\text{(c)}].\end{aligned}$$

This completes the proof of (d) by transfinite induction.
(e) This is also proved by a straightforward transfinite induction on γ.
(f) and (g) Immediate [corollary 6.3.4 and theorem 6.3.5]. □

Example $1 + \omega = \sup_{n<\omega} (1 + n) = \omega < \omega + 1.$

Remark 4 This example shows that ordinal addition (in contrast to cardinal addition) is non-commutative. It is worth stressing just how non-commutative it is: in particular, the analogues for right-addition of α of parts (a), (b), (c), and (g) of proposition 6.3.5 are not generally valid.

Exercises

2. If $\alpha \in \mathbf{On}$, show that $\alpha \geq \omega$ iff $\alpha = 1 + \alpha$.
3. Is it true in general that $\alpha = (\alpha - \beta) + \beta$?
4. Find necessary and sufficient conditions on the ordinals $\alpha, \beta \in \mathbf{On}$ for $\alpha + \beta$ to be a limit ordinal.
5. The mapping $\mathbf{On} \to \mathbf{On}$ given by $\alpha \mapsto \alpha + \beta$ is increasing for all $\beta \in \mathbf{On}$ [proposition 6.3.6(e)]. Show, however, that it is strictly increasing iff β is finite, and normal iff $\beta = 0$.
6. Suppose that $(A, \leq)$ and $(B, \leq)$ are well-ordered collections. Verify that if $f: A \to \boldsymbol{\alpha}$ and $g: B \to \boldsymbol{\beta}$ are isomorphisms then the mapping $h: A \cup B \to \{\gamma : \gamma < \alpha + \beta\}$ given by

$$h(x, i) = \begin{cases} f(x) & \text{if } i = 0 \\ \alpha + g(x) & \text{if } i = 1 \end{cases}$$

is also an isomorphism.

6.4 Ordinal multiplication

Definition

If $(A, \leq)$ and $(B, \leq)$ are two partially ordered collections, we define their *ordered product* AB to be the ordered pair $(A \times B, \leq)$ where $(x_1, y_1) \leq (x_2, y_2)$ iff **either** $y_1 < y_2$ **or** $y_1 = y_2$ and $x_1 \leq x_2$.

This ordering is frequently called the *reverse lexicographic* ordering since it corresponds to the order in which words appear in a Persian dictionary. As with the ordered sum, the ordered product of totally ordered collections is totally ordered. Moreover:

Lemma 6.4.1

The order product of two well ordered collections is well ordered.

Proof If C is a non-empty subcollection of $A \times B$, then from among those elements of C whose B-coordinate is the least possible choose the one whose A-coordinate is the least possible: this is evidently the least element of C. □

Definition

If $\alpha = \text{ord}(A, \leq)$ and $\beta = \text{ord}(B, \leq)$, $\alpha\beta$ is defined to be the ordinal of the ordered product of $(A, \leq)$ with $(B, \leq)$.

Justification As for ordinal addition. □

Remark 1 The meaning of this notation is perhaps best represented by reading $\alpha\beta$ as 'α, β times'. Since this is somewhat unnatural it might seem more appropriate to write $\alpha\beta$ as $\beta\alpha$: indeed Cantor did just this in his first paper on the subject in 1883. He was induced to change his notation in 1887–8 by the observation that the formal properties are thus made much neater; for example, proposition 6.5.6(e) below would in Cantor's original notation take on what he called the 'repulsive' (*abstoßende*) form

$$\alpha^{(\beta+\gamma)} = \alpha^{(\gamma)}\alpha^{(\beta)}.$$

Proposition 6.4.2

$|\alpha\beta| = |\alpha||\beta|$ *for all* $\alpha, \beta \in \mathbf{On}$.

Proof Trivial. □

Proposition 6.4.3

The mapping $\mathbf{On} \times \mathbf{On} \to \mathbf{On}$ *given by* $(\alpha, \beta) \mapsto \alpha\beta$ *satisfies these three equations*:

(*a*) $\alpha 0 = 0$;

(*b*) $\alpha(\beta + 1) = \alpha\beta + \alpha$;

(*c*) $\alpha\lambda = \sup_{\beta<\lambda} \alpha\beta$ *if* λ *is a limit ordinal.*

Proof Straightforward. □

Remark 2 It follows at once from proposition 6.4.3 that for finite ordinals (i.e. natural numbers) this definition coincides with the one given in §4.2.

Corollary 6.4.4

The mapping **On** → **On** *given by* $\beta \mapsto \alpha\beta$ *is normal (strictly normal if* $\alpha > 0$).

Proof Immediate [propositions 6.1.6 and 6.4.3]. □

Theorem 6.4.5 (*Division algorithm*)

If $\alpha, \beta \in$ **On** *and* $\beta \neq 0$, *then there exist unique ordinals* δ *and* $\rho < \beta$ *such that* $\alpha = \beta\delta + \rho$.

Existence If we choose the greatest ordinal δ such that $\beta\delta \leq \alpha$ [corollary 6.4.4 and proposition 6.1.7], then there exists ρ such that $\alpha = \beta\delta + \rho$ [proposition 6.3.6(f)]. Hence

$$\beta\delta + \beta = \beta(\delta + 1) > \alpha = \beta\delta + \rho,$$

and so $\rho < \beta$ by cancellation.

Uniqueness Suppose that $\alpha = \beta\delta + \rho$ as in the theorem but that δ is not the greatest ordinal such that $\beta\delta \leq \alpha$. Then

$$\alpha \geq \beta(\delta + 1) = \beta\delta + \beta > \beta\delta + \rho = \alpha,$$

which is absurd. This proves the uniqueness of δ; the uniqueness of ρ follows from the subtraction algorithm. □

Exercise 1. If $\delta < \tau$ in the above decomposition, show that $\alpha < \beta\tau$.

Remark 3 The division algorithm is particularly useful in the case when $\beta = \omega$, so that $\alpha = \omega\delta + n$ for some $n \in \boldsymbol{\omega}$.

Corollary 6.4.6

α *is a small limit ordinal iff* $\alpha = \omega\beta$ *for some small* $\beta > 0$.

Necessity $\alpha = \omega\beta + n$ for some $\beta \in$ **On** and $n \in \boldsymbol{\omega}$ [division algorithm]. If $n \neq 0$ then $n = m + 1$ for some $m \in \boldsymbol{\omega}$ and so

$$\alpha = \omega\beta + (m + 1) = (\omega\beta + m) + 1,$$

so that α is a successor ordinal, contrary to assumption. Thus $n = 0$ and therefore $\alpha = \omega\beta$ as required.

Sufficiency Suppose that $\omega\beta$ is not zero or a limit. So $\omega\beta = \gamma + 1$ for some $\gamma \in$ **On**. Now $\gamma = \omega\sigma + n$ for some $\sigma \leq \alpha$ and $n \in \boldsymbol{\omega}$. Then

$$\omega\beta = (\omega\sigma + n) + 1 = \omega\sigma + (n + 1).$$

By the uniqueness of the decomposition in the division algorithm, $n + 1 = 0$. Contradiction. □

Proposition 6.4.7

Suppose that $\alpha, \beta, \gamma \in \mathbf{On}$:
(*a*) *if* $\alpha \neq 0$, *then* $\beta < \gamma \Rightarrow \alpha\beta < \alpha\gamma$;
(*b*) *if* $\alpha \neq 0$, *then* $\alpha\beta = \alpha\gamma \Rightarrow \beta = \gamma$;
(*c*) *if* $B \subseteq \mathbf{On}$ *is non-empty and bounded then* $\alpha(\sup_{\beta \in B} \beta) = \sup_{\beta \in B}(\alpha\beta)$;
(*d*) $\alpha(\beta + \gamma) = \alpha\beta + \alpha\gamma$;
(*e*) $\alpha(\beta\gamma) = (\alpha\beta)\gamma$;
(*f*) $\alpha \leq \beta \Rightarrow \alpha\gamma \leq \beta\gamma$.

Proof (a), (b), and (c) Immediate [corollary 6.4.4].
(d) We shall prove this by simple transfinite induction on γ. The case $\gamma = 0$ is trivial. If $\alpha(\beta + \gamma) = \alpha\beta + \alpha\gamma$, then

$$\begin{aligned}\alpha(\beta + (\gamma + 1)) &= \alpha((\beta + \gamma) + 1)\\ &= \alpha(\beta + \gamma) + \alpha\\ &= (\alpha\beta + \alpha\gamma) + \alpha\\ &= \alpha\beta + (\alpha\gamma + \alpha)\\ &= \alpha\beta + \alpha(\gamma + 1).\end{aligned}$$

And if $\lambda \in \mathbf{On}$ is a limit ordinal such that $\alpha(\beta + \gamma) = \alpha\beta + \alpha\gamma$ for all $\gamma < \lambda$, then

$$\alpha(\beta + \lambda) = \sup_{\gamma < \lambda}(\alpha(\beta + \gamma)) = \sup_{\gamma < \lambda}(\alpha\beta + \alpha\gamma) = \alpha\beta + \alpha\lambda.$$

This completes the proof by induction.
(e) and (f) Simple transfinite induction on γ again. □

Remark 4 What was said earlier about the non-commutativity of addition applies equally to multiplication: the analogues for multiplication on the right by α of parts (a) to (d) of proposition 6.4.7 are not valid in general.

Example 1 $(2 + 3)\omega = 5\omega = \omega < \omega 2 = 2\omega + 3\omega$.

Example 2 $(\omega 2 + 1)\omega = \sup_{n<\omega}((\omega 2 + 1)n) = \omega\omega$.

Exercises

2. Show that $\beta\omega$ is a limit ordinal for all $\beta \in \mathbf{On} - \{0\}$ but that not all limit ordinals are of this form.
3. Show that if $\alpha \in \mathbf{On}$ then $\alpha + 1 + \alpha = 1 + \alpha 2$.
4. Find necessary and sufficient conditions on the ordinals $\alpha, \beta \in \mathbf{On}$ for $\alpha\beta$ to be a limit ordinal.
5. (a) If $\alpha, \beta \in \mathbf{On}$, $\alpha \neq 0$, and β is a limit ordinal, show that $(\alpha + 1)\beta = \alpha\beta$.
   (b) The mapping $\mathbf{On} \to \mathbf{On}$ given by $\alpha \mapsto \alpha\beta$ is increasing for all $\beta \in \mathbf{On}$; show that it is strictly increasing iff β is a small successor ordinal.
   (c) Show that if $\beta \in \mathbf{On} - \{0\}$ then β is a limit ordinal iff $n\beta = \beta$ for all $n \in \boldsymbol{\omega} - \{0\}$.

6. Suppose that $(A, \leq)$ and $(B, \leq)$ are well-ordered collections. If $f : A \to \boldsymbol{\alpha}$ and $g : B \to \boldsymbol{\beta}$ are isomorphisms, verify that the mapping $h : A \times B \to \{\gamma : \gamma < \alpha\beta\}$ given by $h(x, y) = \alpha g(y) + f(x)$ is an isomorphism.

6.5 Ordinal exponentiation

Definition

If $(A, \leq)$ and $(B, \leq)$ are totally ordered collections and A has a least element $\perp$, then the *ordered exponential* of $(A, \leq)$ with $(B, \leq)$ is defined to be the ordered pair $({}^{(A)}B, \leq)$ where ${}^{(A)}B$ consists of the functions f from A to B such that $f(x) = \perp$ for all but finitely many $x \in A$, and the ordering is defined so that $f \leq g$ iff **either** $f = g$ **or** $f \neq g$ and $f(x_0) < g(x_0)$, where x_0 is the greatest element of the finite collection $\{x \in A : f(x) \neq g(x)\}$.

The choice of this ordering is determined by our desire to obtain a definition of ordinal exponentiation which obeys the appropriate recursive conditions (see proposition 6.5.3). However, it is undoubtedly less easy to visualize than either the sum or the product.

Lemma 6.5.1

If $(A, \leq)$ and $(B, \leq)$ are well ordered then the ordered exponential $({}^{(A)}B, \leq)$ of A with B is also well ordered.

Proof Suppose that $\mathscr{F}$ is a non-empty subcollection of ${}^{(A)}B$. So it has an element f. Let $a_0, a_1, \ldots, a_{n-1}$ be the elements $a \in A$ such that $f(a) \neq \perp$, arranged so that $a_0 > a_1 > \cdots > a_{n-1}$. Then recursively define elements $b_0, b_1, \ldots, b_{n-1}$ of B so that

$$b_r = \min\{g(a_r) : g \in \mathscr{F} \text{ and } g(a_p) = b_p \text{ for } p < r\}.$$

It is easy to check that if f_0 is given by $f_0(a_r) = b_r$ for $r < n$ and $f_0(x) = \perp$ for $x \in A - \{a_0, a_1, \ldots, a_{n-1}\}$ then f_0 is the least element of $\mathscr{F}$. □

Definition

If $\alpha = \operatorname{ord}(A, \leq)$ and $\beta = \operatorname{ord}(B, \leq)$ then we let $\beta^{(\alpha)}$ denote the ordinal of the ordered exponential $({}^{(A)}B, \leq)$.

Justification As for ordinal addition and multiplication. □

Proposition 6.5.2 (Schönflies 1913)

*If $\alpha, \beta \in$ **On** are both infinite then $|\beta^{(\alpha)}| = \max(|\alpha|, |\beta|)$.*

Proof This follows from the definition and a result on well-orderable cardinals which we shall prove later [corollary 6.7.11]. □

Remark 1 The practice of using results which have not yet been proved is not generally to be recommended: however, you should have no difficulty in checking that on this occasion no logical circle is involved.

Proposition 6.5.3

The mapping **On** $\times$ **On** $\to$ **On** *given by* $(\alpha, \beta) \mapsto \beta^{(\alpha)}$ *satisfies these three equations:*

(*a*) $\beta^{(0)} = 1$

(*b*) $\beta^{(\alpha+1)} = \beta^{(\alpha)}\beta$

(*c*) $\beta^{(\lambda)} = \sup_{\alpha<\lambda} \beta^{(\alpha)}$.

Proof Straightforward. □

Remark 2 Thus the definition coincides with the familiar one when the ordinals in question are finite.

Example 1 $2^{(\omega)} = \sup_{n<\omega} 2^n = \omega$.

Remark 3 One consequence of example 1 worth noting is that

$$|2^{(\omega)}| = |\omega| = \aleph_0 < 2^{\aleph_0} = |2|^{|\omega|}.$$

Thus the exponentiation of ordinals does not mesh neatly with that of cardinals in the way addition and multiplication do (cf. propositions 6.3.2 and 6.4.2). However, this is inevitable so long as we insist on the (surely desirable) requirement that $\alpha \mapsto 2^{(\alpha)}$ be a normal function. If you are nevertheless minded to look for an alternative definition, be warned: if we could define $2^{[\omega]}$ (say) to be a definite ordinal such that $|2^{[\omega]}| = |2|^{|\omega|} = 2^{\aleph_0}$ then there would exist a definite well-ordering on $\mathfrak{P}(\omega)$ (since it definitely has cardinal $2^{\aleph_0}$) and this is something which is known not to be provable from our axioms even if we include the axiom of choice (Feferman 1965).

Corollary 6.5.4

The mapping **On** $\to$ **On** *given by* $\alpha \mapsto \beta^{(\alpha)}$ *is normal if* $\beta > 0$ (*strictly normal if* $\beta > 1$).

Proof Immediate [propositions 6.1.6 and 6.5.3]. □

Theorem 6.5.5 (*Logarithmic algorithm*)

If $\alpha, \beta \in$ **On**, $\alpha > 0$, *and* $\beta > 1$, *then there exist unique* $\gamma, \delta, \rho \in$ **On** *such that* $\alpha = \beta^{(\gamma)}\delta + \rho$ *with* $0 < \delta < \beta$ *and* $\rho < \beta^{(\gamma)}$.

Existence Choose the greatest ordinal γ such that $\beta^{(\gamma)} \leq \alpha$ [corollary 6.5.4 and proposition 6.1.7] and then use the division algorithm to obtain ordinals δ and ρ such that $\rho < \beta^{(\gamma)}$ and $\beta^{(\gamma)}\delta + \rho = \alpha$. If we had $\delta = 0$ we would have $\rho = \alpha \geq \beta^{(\gamma)} > \rho$, which is absurd. And if we had $\delta \geq \beta$ we would have

$$\alpha < \beta^{(\gamma+1)} = \beta^{(\gamma)}\beta \leq \beta^{(\gamma)}\delta \leq \beta^{(\gamma)}\delta + \rho = \alpha,$$

which is also absurd: hence $0 < \delta < \beta$ as required.

Uniqueness Suppose that $\alpha = \beta^{(\gamma)}\delta + \rho$ as in the proposition but γ is not the greatest ordinal such that $\beta^{(\gamma)} \leq \alpha$. Then

$$\begin{aligned}\alpha \geq \beta^{(\gamma+1)} &= \beta^{(\gamma)}\beta \geq \beta^{(\gamma)}(\delta + 1)\\ &= \beta^{(\gamma)}\delta + \beta^{(\gamma)} > \beta^{(\gamma)}\delta + \rho = \alpha,\end{aligned}$$

which is absurd. This proves the uniqueness of γ. The uniqueness of δ and ρ now follows from the division algorithm. □

Exercise

1. If $\gamma < \tau$ in the above decomposition, show that $\alpha < \beta^{(\tau)}$.

Proposition 6.5.6

Suppose that $\alpha, \beta, \gamma \in \mathbf{On}$:
(a) if $\alpha > 1$, then $\beta < \gamma \Leftrightarrow \alpha^{(\beta)} < \alpha^{(\gamma)}$;
(b) if $\alpha > 1$, then $\alpha^{(\beta)} = \alpha^{(\gamma)} \Rightarrow \beta = \gamma$;
(c) if $B \subseteq \mathbf{On}$ is non-empty and bounded, then $\alpha^{(\sup B)} = \sup_{\beta \in B} \alpha^{(\beta)}$;
(d) if $\alpha > 1$, then $\beta \leq \alpha^{(\beta)}$;
(e) $\alpha^{(\beta+\gamma)} = \alpha^{(\beta)}\alpha^{(\gamma)}$;
(f) $(\alpha^{(\beta)})^{(\gamma)} = \alpha^{(\beta\gamma)}$;
(g) $\alpha \leq \beta \Rightarrow \alpha^{(\gamma)} \leq \beta^{(\gamma)}$.

Proof (a), (b), and (c) follow from corollary 6.5.4. (d) follows from (a) and theorem 6.1.2. (e), (f), and (g) may all be proved by simple transfinite induction on γ. □

Exercises

2. Show how to define well-orderings on the *set* $\boldsymbol{\omega}$ whose ordinals are $\omega + 2$, $\omega 2$, $\omega^{(2)}$, and $\omega^{(\omega)}$.
3. Find conditions on the ordinals $\alpha, \beta \in \mathbf{On}$ which are necessary and sufficient to ensure that $\alpha^{(\beta)}$ is a limit ordinal.
4. Show that the mapping $\mathbf{On} \to \mathbf{On}$ given by $\beta \mapsto \beta^{(\alpha)}$ is strictly increasing iff α is a successor ordinal.
5. Show that it is not in general the case that $(\alpha\beta)^{(\gamma)} = \alpha^{(\gamma)}\beta^{(\gamma)}$. [*Hint* Let $\gamma = \omega$.]

Theorem 6.5.7

If $\alpha, \beta \in \mathbf{On}$ and $\beta > 1$, then there exist unique finite sequences $(\gamma_r)_{r \in n}$ and $(\delta_r)_{r \in n}$ of ordinals such that $\gamma_0 > \gamma_1 > \cdots > \gamma_{n-1}$, $0 < \delta_r < \beta$ $(r < n)$ and

$$\alpha = \beta^{(\gamma_0)}\delta_0 + \beta^{(\gamma_1)}\delta_1 + \cdots + \beta^{(\gamma_{n-1})}\delta_{n-1}.$$

Existence Apply the logarithmic algorithm repeatedly: the process must stop after a finite number of steps since otherwise we would obtain a strictly decreasing sequence of ordinals, contradicting the fact that **On** is well ordered.

Uniqueness This follows from the uniqueness of the logarithmic algorithm. □

Remark 4 The expression for α given in theorem 6.5.7 is called the *normal form* of α to the *base* β. Two particular cases are noteworthy: when $\beta = 2$ we obtain the *dyadic normal form*

$$\alpha = 2^{(\gamma_0)} + 2^{(\gamma_1)} + \cdots + 2^{(\gamma_{n-1})};$$

and when $\beta = \omega$ we obtain the so-called *Cantorian normal form*

$$\alpha = \omega^{(\gamma_0)}m_0 + \omega^{(\gamma_1)}m_1 + \cdots + \omega^{(\gamma_{n-1})}m_{n-1}.$$

Example 2

$$\begin{aligned}(\omega 2 + 1)(\omega + 1)3 &= ((\omega 2 + 1)\omega + \omega 2 + 1)3 \\ &= (\omega^{(2)} + \omega 2 + 1)3 \text{ by the example in §6.4} \\ &= \omega^{(2)} + \omega 2 + 1 + \omega^{(2)} + \omega 2 + 1 + \omega^{(2)} + \omega 2 + 1 \\ &= \omega^{(2)} + \omega^{(2)} + \omega^{(2)} + \omega 2 + 1 \\ &= \omega^{(2)}3 + \omega 2 + 1,\end{aligned}$$

which is in Cantorian normal form.

Exercises

6. Find the Cantorian normal form of

$$\omega(\omega + 1)(\omega^{(2)} + 1).$$

7. Find the Cantorian normal form of

$$\omega^{(n+1)} - (1 + \omega + \omega^{(2)} + \cdots + \omega^{(n)}).$$

8. Find the Cantorian normal form of $2^{(\omega\alpha+n)}$, where $\alpha \in \mathbf{On}$ and $n \in \boldsymbol{\omega}$.
9. If $\alpha \in \mathbf{On}$, show that $\alpha = \omega\alpha$ iff $\alpha = \omega^{(\omega)}\beta$ for some $\beta \in \mathbf{On}$. [Consider the Cantorian normal form of α.]
10. Suppose that $(A, \leq)$ and $(B, \leq)$ are well-ordered collections and that $f : A \to \boldsymbol{\alpha}$ and $g : B \to \boldsymbol{\beta}$ are isomorphisms. Suppose that $u \in {}^{(B)}A$: if u is the zero function, let $h(u) = 0$; if not, let $x_0, \ldots, x_{n-1}$ be the elements of B at which u is non-zero, arranged so that $x_0 > x_1 \cdots > x_{n-1}$, and let

$$h(u) = \alpha^{(g(x_0))}f(u(x_0)) + \alpha^{(g(x_1))}f(u(x_1)) + \cdots + \alpha^{(g(x_{n-1}))}f(u(x_{n-1})).$$

Verify that this defines the unique isomorphism $h : {}^{(B)}A \to \{\gamma : \gamma < \alpha^{(\beta)}\}$.
11. An ordinal $\alpha \in \mathbf{On}$ is said to be *indecomposable* if $\beta + \alpha = \alpha$ for all $\beta < \alpha$.
    (a) Show that α is indecomposable iff for any $\beta, \gamma \in \mathbf{On}$ such that $\alpha = \beta + \gamma$ we have either $\alpha = \beta$ or $\alpha = \gamma$.
    (b) Show that if $\beta \neq 0$ then α is indecomposable iff $\beta\alpha$ is.

(c) If α is indecomposable and $0 < \beta < \alpha$ show that $\alpha = \beta\gamma$ for some indecomposable $\gamma \in \mathbf{On}$.
(d) Show that for any $\alpha \in \mathbf{On} - \{0\}$ the least indecomposable ordinal $> \alpha$ is $\alpha\omega$.
(e) Show that a small ordinal is indecomposable iff it is 0 or of the form $\omega^{(\beta)}$ for some $\beta \in \mathbf{On}$.

12. An ordinal $\alpha \in \mathbf{On}$ is said to be *critical* if $\beta\alpha = \alpha$ whenever $0 < \beta < \alpha$.
(a) Show that α is critical iff $\beta\gamma = \alpha \Rightarrow (\beta = \alpha \text{ or } \gamma = \alpha)$.
(b) Show that if $\beta > 1$ then the least critical ordinal $> \beta$ is $\beta^{(\omega)}$.
(c) Deduce that the critical ordinals are 0, 1, 2 and those of the form $\omega^{(\omega^{(\beta)})}$ for $\beta \in \mathbf{On}$.

13. An ordinal $\alpha \in \mathbf{On}$ is called an *epsilon* ordinal if $\beta^{(\alpha)} = \alpha$ whenever $1 < \beta < \alpha$.
(a) Show that α is an epsilon ordinal iff either $\alpha = 1$ or $\beta^{(\gamma)} = \alpha \Rightarrow (\beta = \alpha$ or $\gamma = \alpha)$.
(b) Show that if $\alpha > \omega$ then α is an epsilon ordinal iff $\omega^{(\alpha)} = \alpha$.
(c) Let ε_0 be the least epsilon ordinal $> \omega$. Show that ε_0 exists. What is $|\varepsilon_0|$?

6.6 Goodstein's theorem

When we are dealing with finite ordinals (i.e. natural numbers), a further reduction is possible. If we express m in its normal form to the base n, each of the exponents which appear in this expression must be less than m (since $m < 2^m$); so if we replace each of these exponents by its normal form to the base n, and then repeat the procedure on the exponents thus obtained, and so on, we shall in a finite number of steps arrive at an expression for m (called the *complete normal form* to the base n) in which no number $> n$ appears. If we replace all the occurrences of n in this expression by a variable x we obtain a term which we shall denote $f_{m,n}(x)$.

Remark 1 The method by which the term $f_{m,n}(x)$ is obtained from the natural numbers m and n can be formalized in **PA** and the values of the resulting function $\boldsymbol{\omega} \to \boldsymbol{\omega}$ can be calculated by an explicit mechanical procedure. (In the jargon, $f_{m,n}$ is a 'primitive recursive' function.)

Example 1 $2350 = 3^7 + 3^4.2 + 1 = 3^{3.2+1} + 3^{3+1}.2 + 1$, which is a complete normal form to the base 3. The corresponding term is

$$f_{2350,3}(x) = x^{x.2+1} + x^{x+1}.2 + 1.$$

Example 2 $8192 = 2^{13} = 2^{2^3+2^2+1} = 2^{2^{2+1}+2^2+1}$, which is a complete normal form to the base 2. The corresponding term is

$$f_{8192,2}(x) = x^{x^{x+1}+x^x+1}.$$

Definition

The *Goodstein sequence* $(g_n(m))_{n \geq 1}$ of a natural number m is defined recursively as follows:

$$g_1(m) = m;$$

$$g_{n+1}(m) = \begin{cases} f_{g_n(m),n+1}(n+2) - 1 & \text{if } g_n(m) \neq 0 \\ 0 \quad \text{if } g_n(m) = 0. \end{cases}$$

Expressed in a less compact notation, the first term of the Goodstein sequence for m is m, and the $(n+1)$th is obtained by expressing the nth term in complete normal form to the base $n+1$, changing all the occurrences of $n+1$ to $n+2$, and subtracting 1.

Example 3 The Goodstein sequence of 3 starts as follows:

$$g_1(3) = 3 = 2 + 1;$$

$$g_2(3) = 3;$$

$$g_3(3) = 4 - 1 = 3;$$

$$g_4(3) = 2;$$

$$g_5(3) = 1;$$

$$g_6(3) = 0.$$

Example 4 The Goodstein sequence of 51 starts as follows:

$$g_1(51) = 51 = 2^{2^2+1} + 2^{2^2} + 2 + 1;$$

$$g_2(51) = 3^{3^3+1} + 3^{3^3} + 3 \sim 10^{13};$$

$$g_3(51) = 4^{4^4+1} + 4^{4^4} + 3 \sim 10^{155};$$

$$g_4(51) = 5^{5^5+1} + 5^{5^5} + 2 \sim 10^{2185};$$

$$g_5(51) = 6^{6^6+1} + 6^{6^6} + 1 \sim 10^{36\,306};$$

$$g_6(51) = 7^{7^7+1} + 7^{7^7} \sim 10^{695\,975};$$

$$g_7(51) = 8^{8^8+1} + 8^{8^8} - 1 \sim 10^{15\,151\,337};$$

and so on.

The Goodstein sequence of 3 reaches 0 in 6 steps. The Goodstein sequence of 4 also reaches 0, but it takes approximately $10^{121\,210\,700}$ steps to do so. We intend now to use the theory of ordinals to show that every Goodstein sequence reaches 0 eventually.

If we replace x in the term $f_{m,n}(x)$ by an ordinal $\alpha < \varepsilon_0$ and evaluate the ordinal expression thus obtained, we obtain another ordinal $f_{m,n}(\alpha) < \varepsilon_0$. By this means we can define a function $\boldsymbol{\varepsilon}_0 \to \boldsymbol{\varepsilon}_0$ which extends the function $\boldsymbol{\omega} \to \boldsymbol{\omega}$ we defined previously.

Definition

The *transfinite Goodstein sequence* $(G_n(m))_{n \geq 1}$ of a natural number m is defined by the equation

$$G_n(m) = \begin{cases} f_{g_n(m),n+1}(\omega) & \text{if } g_n(m) \neq 0 \\ 0 & \text{if } g_n(m) = 0. \end{cases}$$

So the transfinite Goodstein sequence is obtained from the Goodstein sequence by changing all the occurrences of $n + 1$ into ω rather than $n + 2$.

Remark 2 It would be tedious but straightforward to formalize the foregoing precisely in **GA**.

Example 5 The transfinite Goodstein sequence of 3 starts as follows:

$$G_1(3) = \omega + 1;$$
$$G_2(3) = \omega;$$
$$G_3(3) = 3;$$
$$G_4(3) = 2;$$
$$G_5(3) = 1;$$
$$G_6(3) = 0.$$

Example 6 The transfinite Goodstein sequence of 51 starts as follows:

$$G_1(51) = \omega^{\omega^\omega+1} + \omega^{\omega^\omega} + \omega + 1;$$
$$G_2(51) = \omega^{\omega^\omega+1} + \omega^{\omega^\omega} + \omega;$$
$$G_3(51) = \omega^{\omega^\omega+1} + \omega^{\omega^\omega} + 3;$$
$$G_4(51) = \omega^{\omega^\omega+1} + \omega^{\omega^\omega} + 2;$$
$$G_5(51) = \omega^{\omega^\omega+1} + \omega^{\omega^\omega} + 1;$$
$$G_6(51) = \omega^{\omega^\omega+1} + \omega^{\omega^\omega};$$

and so on.

Theorem 6.6.1 (Goodstein 1944)
$(\forall m \in \boldsymbol{\omega})(\exists n \in \boldsymbol{\omega})(g_n(m) = 0)$.

Proof If m is such that $g_n(m) \neq 0$ for all $n \in \boldsymbol{\omega}$, then $(G_n(m))_{n\geq 1}$ is a strictly decreasing sequence of ordinals, which is absurd. □

Goodstein's theorem is an assertion about the arithmetic of the natural numbers. We have just shown that it has a very short proof using the theory of ordinals. One might expect it also to have an elementary arithmetical proof—albeit perhaps a much longer one. But it does not: it has been shown (Kirby and Paris 1982) that Goodstein's theorem cannot be proved in **PA** (although it can of course be stated in the language of **PA**).

Need a number-theorist believe Goodstein's theorem? Let us consider the position of a number-theorist who is familiar with the counting numbers and the arithmetical operations upon them, and believes that Peano's axioms are all true when interpreted as assertions about them, but has no views either way about the existence of the infinite sets which we have posited here. (We shall assume for convenience that this number-theorist is male and believes in some suitable sense in the reality of numbers, but nothing in what follows hinges on either assumption.) Can we persuade him to believe that all the arithmetical assertions which we can prove set-theoretically about $\boldsymbol{\omega}$ are true about the counting numbers he knows and loves?

Notice straightaway that if the axioms of **PA** are complete (i.e. strong enough to determine the truth or falsity of every arithmetical proposition) and **GA** is consistent (as I hope and pray it is) then he has no choice in the matter. To see this, suppose that Φ is an arithmetical proposition and $\Phi^{(\omega)}$ is (by a slight abuse of our previous notation) the corresponding set-theoretic claim about the structure $\boldsymbol{\omega}$. Certainly if Φ is provable arithmetically (i.e. in **PA**) then $\Phi^{(\omega)}$ is provable in **GA**; and conversely if $\Phi^{(\omega)}$ is provable in **GA** then (not $\Phi^{(\omega)}$) is not provable (since **GA** is formally consistent). Hence (not Φ) is not provable in **PA**. So Φ *is* provable in **PA** (since we are assuming that **PA** is complete). All that remains, then, in order to legitimize the use of set-theoretic arguments in number theory is the task of showing that **PA** is complete and **GA** is consistent: this task is known as 'Hilbert's programme'.

But **PA** is *not* complete. There are explicit arithmetical assertions which are true in some models and false in others: Goodstein's theorem is an example. This is unquestionably very surprising at first: it seems to fight against our earlier claim [corollary 4.1.5] that all Dedekind algebras are isomorphic, since it plainly follows from this that the same arithmetical sentences are true in all Dedekind algebras. But there is no contradiction: contrary to the impression we may have given earlier, it is not the case that every model of **PA** is a Dedekind algebra (although the converse certainly holds).

The ambiguity lies in what we are prepared to count as 'properties' for the purpose of the induction rule. The definition of $\boldsymbol{\omega}$ interprets this rule by requiring that $\boldsymbol{\omega}$ should have no proper s-closed subsets containing 0. But the only properties which the number-theorist seems obliged to count are those which are expressible in his specification language, i.e. arithmetical ones. In set-theoretic terms this amounts roughly to requiring only that $\boldsymbol{\omega}$ should have no proper *recursive* s-closed subsets containing 0. It is the impredicativity in the AXIOM OF SUBCOLLECTIONS in combination with the axiom of infinity which makes this ambiguity possible. Of course, from the point of view of the set-theorist it would be perverse to restrict the induction principle to arithmetical properties, but there is no reason for the number-theorist to share this view unless he can be persuaded of the reality of the infinite sets referred to in the proofs in question. Because of the incompleteness in his specification, it is simply not enough for us to convince him of the formal consistency of set theory. Of course, it may be that the number-theorist will be able to refine his specification to the point when it can decide Goodstein's theorem. (He could, for example, convince himself on intuitive grounds that Goodstein's theorem itself should be added to the list of axioms. However, it is not at all clear what form this intuition might take or how he should go about obtaining it: Isaacson (1987) has suggested that there is no way of perceiving in purely arithmetic terms the truth of any arithmetical proposition—such as Goodstein's theorem—which is not provable from Peano's axioms.) In any case this would merely postpone the problem, since any specification for number theory stronger than **PA** which satisfies the requirements for mechanical checking which we mentioned in the Introduction is incomplete (Gödel 1931).

Nor is the matter immediately settled for a liberated number-theorist who accepts the reality of infinite sets of numbers. For there is still then a choice to be made about what *sort* of infinite sets of numbers exist. In particular, it is not impossible that there could be a (consistent) variant of set theory in which Goodstein's theorem is provably false. (Of course, the conception of infinite sets which motivated such a theory would have to be very different from the one we have outlined in this book: it would have to be a theory in which no coherent theory of ordinals could be developed and would therefore presumably be highly non-well-founded.) This phenomenon—alternative views about the behaviour of infinite sets ('the ideal part') resulting in different things being provable about numbers ('the real part')—arises even more pressingly in the case of the continuum hypothesis (see §7.5), since in this case there is not even a consensus among mathematicians about whether to accept it (unlike the situation with ordinals).

Let us mention here in contrast the attitude of affine geometers towards 'lines at infinity': projective methods can provide us with more elegant, more intuitive, or more concise proofs of affine theorems, but they do not permit us to prove things which were otherwise unprovable. The affine geometers are therefore free to regard lines at infinity as no more than a technical device and need take no view about their existence or reality once the formal consistency of the method has been established. (*Projective* geometers will of course see the matter somewhat differently.) Similar remarks apply to certain uses of complex numbers in the theory of real polynomials (and, as we remarked in Chapter 3, to the use of proper classes in Zermelo-type set theory). But the number-theorist must take a different view about infinite sets: the assumptions he makes about sets affect what is provable about numbers.

It may seem odd that a book on set theory should contain an attempt to undermine the reader's belief in its usefulness. But this is not (quite) my intention. What I want to make clear is that in order to believe a result whose proof uses set-theoretic methods one must believe something about the conception of sets on which those methods are based. It seems to me to be almost unavoidable that the thing one must believe is a form of realism: certainly, formal consistency is not enough—Gödel's incompleteness theorem tells us that—and it is difficult to see how any kind of nominalism could be enough either.

6.7 Well-orderable cardinals

The question whether a collection is well-orderable depends only on its cardinality. With this in mind we define a cardinal $\mathfrak{a} = \text{card}(A)$ (by a minor abuse of language) to be *well-orderable* if A is well-orderable; this definition is therefore independent of the choice of representative collection A. The class of all well-orderable small cardinals will be denoted **Wo**. In this section we shall show that the arithmetic of infinite well-orderable cardinals is particularly simple: addition and multiplication are completely trivial; only exponentiation remains of any interest.

Example 1 If $\mathfrak{a} \leq \mathfrak{b}$ and $\mathfrak{b}$ is well-orderable, then $\mathfrak{a}$ is well-orderable. Hence in particular every cardinal $\leq \aleph_0$ belongs to **Wo**.

Remark 1 The well-orderable cardinals are precisely those of the form $|\alpha|$ for some ordinal α.

Remark 2 If λ is the ordinal of the class $\mathbf{Wo} - \boldsymbol{\omega}$ of small infinite well-orderable cardinals, then there is a unique isomorphism from $\boldsymbol{\lambda}$ onto $\mathbf{Wo} - \boldsymbol{\omega}$: its graph is denoted $\alpha \mapsto \aleph_\alpha$; the least element of the set $\{\beta \in \mathbf{On} : |\beta| = \aleph_\alpha\}$ is denoted ω_α. $\aleph$ is the Hebrew letter 'aleph': because of this the elements of $\mathbf{Wo} - \boldsymbol{\omega}$ (i.e. the small infinite well-orderable cardinals) are commonly called *alephs*. Note that λ is a limit ordinal $\leq$ On; indeed if we assume the AXIOM OF REPLACEMENT, $\lambda =$ On and so $\aleph_\alpha$ exists for all $\alpha \in \mathbf{On}$.

Example 2 The least infinite cardinal is therefore $\aleph_0$ and the least infinite ordinal is ω_0. This is of course consistent with our previous usage.

Example 3 The least uncountable well-orderable cardinal is $\aleph_1$ and the least uncountable ordinal is ω_1.

Proposition 6.7.1

$(\mathbf{Wo}, \leq)$ *is a well ordered class.*

roof This follows at once from the fact that $(\mathbf{On}, \leq)$ is well ordered and the mapping $\mathbf{On} \to \mathbf{Wo}$ given by $\alpha \mapsto |\alpha|$ is surjective and increasing. □

emark 3 In particular, any two well-orderable cardinals are comparable: therefore, every well-orderable cardinal is either finite or infinite.

roposition 6.7.2

If $\mathfrak{a} \in \mathbf{Wo}$ *is infinite, then* $2\mathfrak{a} = \mathfrak{a}$.

roof We shall prove by transfinite induction that $2|\alpha| = |\alpha|$ for every small infinite ordinal α. This is certainly true for $\alpha = \omega$ since $2\aleph_0 = \aleph_0$. If it is true for α, then

$$|2(\alpha + 1)| = |2\alpha + 2| = 2|\alpha| + 2$$
$$= |\alpha| + 2 = |\alpha| + 1 = |\alpha + 1|,$$

and so it is true for $\alpha + 1$. Finally, if λ is a limit ordinal and the hypothesis is true for $\omega \leq \alpha < \lambda$, then $\lambda = \omega\beta$ for some $\beta < \lambda$ [corollary 6.4.6], so that

$$|2\lambda| = |2\omega\beta| = 2\aleph_0|\beta| = \aleph_0|\beta| = |\omega\beta| = |\lambda|,$$

and hence it is true for λ. This completes the proof. □

orollary 6.7.3

If $\mathfrak{a}, \mathfrak{b} \in \mathbf{Wo}$ *are infinite, then* $\mathfrak{a} + \mathfrak{b} = \max(\mathfrak{a}, \mathfrak{b})$.

roof Either $\mathfrak{a} \leq \mathfrak{b}$ or $\mathfrak{b} \leq \mathfrak{a}$ [proposition 6.7.1]: suppose for the sake of argument that $\mathfrak{a} \leq \mathfrak{b}$. Then

$$\mathfrak{b} \leq \mathfrak{a} + \mathfrak{b} \leq \mathfrak{b} + \mathfrak{b} = 2\mathfrak{b} = \mathfrak{b} \quad \text{[proposition 6.7.2]},$$

and so $\mathfrak{a} + \mathfrak{b} = \mathfrak{b} = \max(\mathfrak{a}, \mathfrak{b})$. □

roposition 6.7.4

If $\mathfrak{a} \in \mathbf{Wo}$ *is infinite, then* $\mathfrak{a}^2 = \mathfrak{a}$.

roof We shall prove by transfinite induction that $|\alpha|^2 = |\alpha|$ for every small infinite ordinal α. This is certainly true for $\alpha = \omega$ since $\aleph_0{}^2 = \aleph_0$. If it is true for α, then

$$|\alpha + 1|^2 = (|\alpha| + 1)^2 = |\alpha|^2 = |\alpha| = |\alpha| + 1 = |\alpha + 1|;$$

so it is true for $\alpha + 1$. If λ is a small limit ordinal such that the result is true for $\omega \leq \alpha < \lambda$, then [corollary 6.4.6] there are two possibilities to consider:

(I) if $\lambda = \omega n$ for some $n \in \boldsymbol{\omega} - \{0\}$, then

$$|\lambda|^2 = |\omega n|^2 = n^2\aleph_0{}^2 = n\aleph_0 = |\omega n| = |\lambda|;$$

(II) if $\lambda = \omega\alpha$ with $\omega \le \alpha < \lambda$, then

$$|\lambda|^2 = |\omega\alpha|^2 = \aleph_0{}^2|\alpha|^2 = \aleph_0|\alpha| = |\omega\alpha| = |\lambda|.$$

Thus in either case the result is true for λ. This completes the proof. □

Corollary 6.7.5

If $\mathfrak{a}, \mathfrak{b} \in \mathbf{Wo}$ *are infinite, then* $\mathfrak{a}\mathfrak{b} = \max(\mathfrak{a}, \mathfrak{b})$.

Proof Suppose for the sake of argument that $\mathfrak{a} \le \mathfrak{b}$. Then

$$\mathfrak{b} \le \mathfrak{a}\mathfrak{b} \le \mathfrak{b}\mathfrak{b} = \mathfrak{b}^2 = \mathfrak{b} \quad \text{[proposition 6.7.4]},$$

so that $\mathfrak{a}\mathfrak{b} = \mathfrak{b} = \max(\mathfrak{a}, \mathfrak{b})$. □

Theorem 6.7.6 (Hartogs 1915)

If $\mathfrak{a}$ *is a cardinal then there is a well-orderable cardinal* $\mathfrak{b}$ *such that* $\mathfrak{b} \not\le \mathfrak{a}$

Proof Suppose on the contrary that $\mathfrak{a}$ is an upper bound for the well-orderabl cardinals. Then it is easy to show (see exercise 3 of §6.2) that every ordina is contained in the fourth day after $D(\mathfrak{a})$. Hence $\{\alpha : \alpha$ is an ordinal exists [AXIOM OF SUBCOLLECTIONS], contradicting Burali-Forti' paradox. □

Definition

If $\mathfrak{a} \in \mathbf{Cn}$ then we let $\mathfrak{a}^+$ denote the least $\mathfrak{b} \in \mathbf{Wo}$ such that $\mathfrak{b} \not\le \mathfrak{a}$ *an* $\mathfrak{b} < 2^{2^{\mathfrak{a}^2}}$.

Justification Let $\mathfrak{a}^+$ be the least well-orderable cardinal $\not\le \mathfrak{a}$ [Hartogs theorem]. Let A be a set such that $\text{card}(A) = \mathfrak{a}$ and let α be the leas ordinal such that $|\alpha| = \mathfrak{a}^+$. It is easy to check that the mapping $f : \mathfrak{P}(\boldsymbol{\alpha}) \to$ $\mathfrak{P}(\mathfrak{P}(A \times A))$ given by

$$f(X) = \{r \subseteq A \times A : \text{ord}(\text{dom}[r], r) \in X\}$$

is injective. So

$$\mathfrak{a}^+ = |\alpha| < \text{card}(\mathfrak{P}(\boldsymbol{\alpha})) \le \text{card}(\mathfrak{P}(\mathfrak{P}(A \times A))) = 2^{2^{\mathfrak{a}^2}}.$$ □

Proposition 6.7.7

$\mathbf{Wo}$ *is a proper class which is unbounded in* $\mathbf{Cn}$.

Proof $\mathbf{Wo}$ is unbounded in $\mathbf{Cn}$: it follows that it is not a set [proposition 5.2.7] □

Theorem 6.7.8

If A *is an infinite well-orderable set, then* $\text{card}(\mathfrak{F}(A)) = \text{card}(A)$.

Proof Suppose not. Then there exists an infinite ordinal $\alpha \in \mathbf{On}$ such tha $\text{card}(\mathfrak{F}(\boldsymbol{\alpha})) \ne |\alpha|$: choose α as small as possible. (Note that α is therefor the least element of $\{\beta : |\beta| = |\alpha|\}$.) Now if $X \in \mathfrak{F}(\boldsymbol{\alpha})$ then we can let $X = \{\gamma_0, \gamma_1, \ldots, \gamma_{n-1}\}$ with $\gamma_0 > \gamma_1 > \cdots > \gamma_{n-1}$ and define

$$\mathrm{f}(X) = 2^{(\gamma_0)} + 2^{(\gamma_1)} + \cdots + 2^{(\gamma_{n-1})}$$

(unless $X = \varnothing$, in which case let $\mathrm{f}(X) = 0$). Now if $0 \leq r \leq n-1$ then either γ_r is finite, in which case

$$|2^{(\gamma_r)}| < |\omega| \leq |\alpha|,$$

or γ_r is infinite, in which case

$$\begin{aligned} |2^{(\gamma_r)}| &= \mathrm{card}(\mathfrak{F}(\boldsymbol{\gamma}_r)) \quad \text{[proposition 5.1.3(b)]} \\ &= |\gamma_r| \quad \text{by the induction hypothesis} \\ &< |\alpha| \quad \text{since } \gamma_r < \alpha \text{ and } \alpha \text{ is the least element of } \{\beta : |\beta| = |\alpha|\}. \end{aligned}$$

So

$$\begin{aligned} |\mathrm{f}(X)| &= \sum_{r \in \boldsymbol{n}} |2^{(\gamma_r)}| \quad \text{[proposition 6.3.1]}, \\ &< |\alpha| \quad \text{[corollary 6.7.3]}, \end{aligned}$$

and therefore $\mathrm{f}(X) < \alpha$. In other words, we have defined a mapping $\mathrm{f} : \mathfrak{F}(\boldsymbol{\alpha}) \to \boldsymbol{\alpha}$; moreover, this mapping is bijective [theorem 6.5.7]. It follows that $\mathrm{card}(\mathfrak{F}(\boldsymbol{\alpha})) = |\alpha|$. Contradiction. □

Proposition 6.7.9

If A is an infinite well-orderable set, then $\mathrm{card}(\mathrm{String}(A)) = \mathrm{card}(A)$.

Proof Each element of $\mathrm{String}(A)$ is a function from $\boldsymbol{n}$ to A, hence a finite subset of $\boldsymbol{\omega} \times A$. So $\mathrm{String}(A) \subseteq \mathfrak{F}(\boldsymbol{\omega} \times A)$. Now $\boldsymbol{\omega}$ and A are both well-orderable, hence so is $\boldsymbol{\omega} \times A$ [lemma 6.4.1]. Therefore

$$\begin{aligned} \mathrm{card}(\mathrm{String}(A)) &\leq \mathrm{card}(\mathfrak{F}(\boldsymbol{\omega} \times A)) \\ &= \mathrm{card}(\boldsymbol{\omega} \times A) \quad \text{[theorem 6.7.8]} \\ &= \aleph_0\, \mathrm{card}(A) \\ &= \mathrm{card}(A) \quad \text{[corollary 6.7.5]}. \end{aligned}$$

The result follows, since the inequality $\mathrm{card}(A) \leq \mathrm{card}(\mathrm{String}(A))$ is obvious. □

Proposition 6.7.10

If $(A, \leq)$ and $(B, \leq)$ are infinite well-orderable sets then $\mathrm{card}(\mathfrak{F}(A, B)) = \max(\mathrm{card}(A), \mathrm{card}(B))$.

Proof If $f \in \mathfrak{F}(A, B)$, let $\{x_0, \ldots, x_{n-1}\}$ be the domain of f arranged in increasing order with respect to some well-ordering on A, and let

$$g(f) = (\{x_0, \ldots, x_{n-1}\}, (f(x_r))_{r \in n}).$$

The mapping $g : \mathfrak{F}(A, B) \to \mathfrak{F}(A) \times \mathrm{String}(B)$ thus defined is evidentl injective. Hence

$$\begin{aligned} \mathrm{card}(\mathfrak{F}(A, B)) &\leq \mathrm{card}(\mathfrak{F}(A) \times \mathrm{String}(B)) \\ &= \mathrm{card}(\mathfrak{F}(A))\,\mathrm{card}(\mathrm{String}(B)) \\ &= \mathrm{card}(A)\,\mathrm{card}(B) \quad [\text{theorem 6.7.8 and proposition 6.7.9}] \\ &= \max(\mathrm{card}(A), \mathrm{card}(B)) \quad [\text{corollary 6.7.5}]. \end{aligned}$$

The converse inequality is obvious, whence the result. □

Corollary 6.7.11

If $(A, \leq)$ and $(B, \leq)$ are infinite well ordered collections then

$$\mathrm{card}({}^{(A)}B) = \max(\mathrm{card}(A), \mathrm{card}(B)).$$

Proof If $\bot$ is the least element of B then there is an obvious one-to-one correspondence between ${}^{(A)}B$ and $\mathfrak{F}(A, B - \{\bot\})$ and so

$$\begin{aligned} \mathrm{card}({}^{(A)}B) &= \mathrm{card}(\mathfrak{F}(A, B - \{\bot\})) \\ &= \max(\mathrm{card}(A), \mathrm{card}(B - \{\bot\})) \quad [\text{proposition 6.7.10}] \\ &= \max(\mathrm{card}(A), \mathrm{card}(B)) \quad \text{since } B \text{ is infinite.} \end{aligned}$$ □

Exercises

1. If $\mathfrak{b} \in \mathbf{Wo}$ is infinite and $\mathfrak{a} \in \mathbf{Cn}$, show that $\mathfrak{a} = 2^{\mathfrak{b}}$ iff $\mathfrak{a} \geq \mathfrak{b}$ and $\mathfrak{a} + \mathfrak{b} = 2^{\mathfrak{b}}$.
2. If $\mathfrak{a} = \mathrm{card}(A)$ and $\mathfrak{b} = \mathrm{card}(B)$, let us write $\mathfrak{a} \leq * \mathfrak{b}$ if either $A = \varnothing$ or there exists a surjective mapping $B \to A$. Establish the following results.
   (a) $\mathfrak{a} \leq \mathfrak{b} \Rightarrow \mathfrak{a} \leq * \mathfrak{b}$.
   (b) The converse holds if $\mathfrak{b}$ is well-orderable.
   (c) $\mathfrak{a} \leq * \mathfrak{b} \Rightarrow 2^{\mathfrak{a}} \leq 2^{\mathfrak{b}}$.
   (d) $\aleph_{\alpha+1} \leq * 2^{\aleph_\alpha}$.
   (e) $\aleph_{\alpha+1} < 2^{2^{\aleph_\alpha}}$.
   (f) If $(A_i)_{i \in I}$ is a family of sets, then
   $$\mathrm{card}\left(\bigcup_{i \in I} A_i\right) \leq * \mathrm{card}\left(\dot{\bigcup_{i \in I}} A_i\right).$$
3. (a) Given $\mathfrak{b} \in \mathbf{Wo} - \{0\}$, find an infinite $\mathfrak{a} \in \mathbf{Cn}$ such that $\mathfrak{a}^{\mathfrak{b}} = \mathfrak{a}$.
   (b) Can we choose $\mathfrak{a} \leq \mathfrak{b}$?
4. If A is an infinite well-orderable set of cardinal $\mathfrak{a}$, show that each of the following sets has cardinal $2^{\mathfrak{a}}$:
   (a) the set of infinite subsets of A;
   (b) the set of subsets of A equinumerous with A;
   (c) the set of equivalence relations on A;
   (d) the set of well-orderings on A.

7 The axiom of choice

We return now to the consideration of the conception of set theory which our axiom system is intended to codify. We have already (in the note on 'Indefinable collections' in §1.2) remarked on the question whether we should allow the possibility of collections which are not, even in principle, definable. We left this question unresolved then, but in §5.5 we were forced, in proving that every set is finite or infinite, to assume the existence of a sequence which we could not explicitly define. The axiom which we used there—that of countable choice—is the prototype for the ones we shall be considering here. In particular, the assertion which we obtain by removing from it the assumption of countability—the so-called 'axiom of choice'—is central in what follows.

7.1 Choice functions and inductively ordered collections

Definition

A partially ordered collection $(A, \leq)$ is said to be *inductively ordered* (and $\leq$ is said to be an *inductive ordering* on A) if every totally ordered subcollection of A has a supremum in A.

Remark 1 Every inductively ordered collection is non-empty and has a least element $\perp$ (since $\varnothing$ is totally ordered).

Remark 2 For a collection $\mathscr{A}$ to be inductively ordered by inclusion it is sufficient (but not necessary) that $\bigcup \mathscr{B} \in \mathscr{A}$ for every chain $\mathscr{B} \subseteq \mathscr{A}$.

Example 1 Every complete partially ordered collection is inductively ordered; so is every finite one.

Example 2 A totally ordered collection is inductive iff it is complete.

Exercises

1. Show that the set $\mathfrak{P}(A, B)$ of functions f such that $\mathrm{dom}[f] \subseteq A$ and $\mathrm{im}[f] \subseteq B$ is inductively ordered by inclusion.
2. If $(A, \leq)$ is inductively ordered and $a \in A$, then $\uparrow a$ is also inductively ordered.
3. Is $\mathfrak{F}(\omega)$ inductively ordered by inclusion?
4. Let Well(A) denote the collection of all relations on A which are well-orderings. If $r, r' \in$ Well(A), define $r \leq r'$ iff $r \subseteq r'$ and $\mathrm{dom}[r]$ is an initial subcollection

of the well ordered collection (dom$[r']$, r'). Show that (Well(A), $\leq$) is an indu tively ordered collection, but that Well(A) need not be inductively ordered b inclusion.

Lemma 7.1.1 (Bourbaki 1949a)

If $(A, \leq)$ is inductively ordered, then every mapping $f: A \to A$ such th $f(x) \geq x$ for all $x \in A$ has a definite fixed point.

Proof Suppose that $f: A \to A$ is such a mapping and let α be the least ordin such that $|\alpha| \nleq \text{card}(A)$ [Hartogs' theorem]. Then [simple principle transfinite recursion] there exists a unique mapping $g: \boldsymbol{\alpha} \to A$ such th

$$g(0) = \bot;$$

$$g(\beta^+) = f(g(\beta)) \text{ if } \beta^+ < \alpha;$$

$$g(\lambda) = \sup g[\boldsymbol{\lambda}] \text{ for every limit ordinal } \lambda < \alpha.$$

Evidently g is normal [proposition 6.1.6], but it is not strictly norm (since if it were it would in particular be one-to-one and we would hav $|\alpha| \leq \text{card}(A)$). So there exists $\beta < \alpha$ such that $g(\beta^+) = g(\beta)$ [propositio 6.1.6]. If we choose the least such β (to be definite) and let $b = g(\beta)$ the

$$f(b) = f(g(\beta)) = g(\beta^+) = g(\beta) = b. \qquad \Box$$

Remark 3 Let us say that A is *quasi-inductively ordered* by $\leq$ if every totally ordere subcollection has an upper bound (not necessarily a least one) in A. Th analogue of Bourbaki's lemma for quasi-inductively ordered sets is no provable in basic set theory and indeed is equivalent to the axiom choice (§7.3).

Exercises

5. Show by an example that a function of the kind referred to in lemma 7.1.1 nee not have a least fixed point.
6. Show that if $(A, \leq)$ is quasi-inductively ordered and B is a cofinal subcollectio of A then B is quasi-inductively ordered by the inherited partial ordering.

Theorem 7.1.2

If $(A, \leq)$ is a well-orderable inductively ordered collection, then $(A, \leq)$ ha a maximal element.

Proof Suppose that A has no maximal element. Then for each $x \in A$ there exis elements $y \in A$ such that $y > x$. So if we choose a well-ordering on A w can define a function $g: A \to A$ by letting $g(x)$ be the least such y wit respect to the well-ordering of A which we have chosen. Evidently $g(x) > x$ for all $x \in A$, contradicting lemma 7.1.1. $\Box$

Warning The well-ordering referred to in the proof of theorem 7.1.2 wil not in general be the same as the given inductive ordering on A: distinguis carefully between 'well ordered' and 'well-orderable'.

The example of an inductive ordering which we gave in exercise 4 is not typical: almost all of the inductive orderings we shall come across will be cases of the inclusion relation. In particular, the following more restrictive condition occurs frequently (especially in algebra).

Definition

A collection $\mathscr{A}$ is said to have *finite character* if it satisfies these two conditions:

(1) if $\mathscr{B}$ is a non-empty subcollection of $\mathscr{A}$ then $\bigcap\mathscr{B} \in \mathscr{A}$;
(2) if C is a collection such that for every finite subcollection F of C there exists $A_F \in \mathscr{A}$ such that $F \subseteq A_F$, then there exists $A \in \mathscr{A}$ such that $C \subseteq A$.

Remark 4 It follows from the first condition that every collection $\mathscr{A}$ of finite character is boundedly complete and (if it is non-empty) has a least element with respect to inclusion. So every subcollection of $\mathscr{A}$ which is bounded above in $\mathscr{A}$ has a supremum in $\mathscr{A}$.

Remark 5 A collection $\mathscr{A}$ is both hereditary and of finite character iff it satisfies the following condition:

A belongs to $\mathscr{A}$ iff every finite subcollection of A belongs to $\mathscr{A}$.

Many authors use the expression 'finite character' to describe collections satisfying this condition.

Exercise

7. Let us say $\mathscr{A}$ is of *quasi-finite character* if it satisfies the second condition in the definition of finite character. Show that if $\mathscr{A}$ is non-empty and of quasi-finite character then it is quasi-inductively ordered by inclusion.

Example 3 Somewhat trivially, $\varnothing$ is of finite character.

Example 4 The collection $\mathrm{Po}(A)$ of partial orderings on a collection A is of finite character but not in general hereditary.

Example 5 If A is partially ordered by $\leq$ then the collection of all subcollections of A which are totally ordered by the inherited partial ordering is hereditary and of finite character.

Proposition 7.1.3

If $\mathscr{A}$ is a non-empty collection of finite character, then it is inductively ordered by inclusion.

Proof Let $\mathscr{B}$ be a chain in $\mathscr{A}$ and let $B = \bigcup\mathscr{B}$. Now certainly $\varnothing$ is contained in an element of $\mathscr{A}$ since $\mathscr{A}$ is non-empty. So consider a non-empty finite subcollection $\{b_0, \ldots, b_{n-1}\}$ of B. For $0 \leq r \leq n-1$ there exists $B_r \in \mathscr{B}$ such that $b_r \in B_r$. Now $\{B_0, B_1, \ldots, B_{n-1}\}$ is a finite non-empty chain and therefore has a greatest element B_{r_0} [theorem 4.4.1]. So $\{b_0, b_1, \ldots, b_{n-1}\}$ $B_{r_0} \in \mathscr{B} \subseteq \mathscr{A}$, i.e. $\{b_0, b_1, \ldots, b_{n-1}\}$ is contained in an element of $\mathscr{A}$. Wh we have now shown is that every finite subcollection of B is contained an element of $\mathscr{A}$. So B itself is contained in an element of $\mathscr{A}$, i.e. $\mathscr{B}$ bounded above and therefore has a supremum in $\mathscr{A}$. [

Definition

A *choice function* is a function f such that $f(A) \in A$ for all $A \in \mathrm{dom}[f$ The collection of all choice functions f such that $\mathrm{dom}[f] \subseteq \mathscr{A}$ is denot choice($\mathscr{A}$).

If f is a choice function, then clearly $\varnothing \notin \mathrm{dom}[f]$; if $\mathrm{dom}[f]$ $\mathscr{A} - \{\varnothing\}$ we shall say that f is a choice function for $\mathscr{A}$.

Lemma 7.1.4

Suppose that $\mathscr{A}$ is a collection.

(*a*) choice($\mathscr{A}$) *is a hereditary collection of finite character.*

(*b*) *The maximal elements of* choice($\mathscr{A}$) *are the choice functions for $\mathscr{A}$.*

Proof (a) Trivial.

(b) Let $f \in$ choice($\mathscr{A}$). Then clearly $\varnothing \notin \mathrm{dom}[f]$, and so $\mathrm{dom}[f]$ $\mathscr{A} - \{\varnothing\}$. But if $\mathrm{dom}[f] \subset \mathscr{A} - \{\varnothing\}$, then there exists B $(\mathscr{A} - \{\varnothing\}) - \mathrm{dom}[f]$: if $x \in B$, then $f \cup \{(B, x)\}$ is a choice functio which strictly contains f. So f is maximal in choice($\mathscr{A}$) iff $\mathrm{dom}[f]$ $\mathscr{A} - \{\varnothing\}$. [

Theorem 7.1.5 (Zermelo 1904)

For every well-ordering on a collection A there exists a definite choi function for $\mathfrak{P}(A)$, and conversely.

Necessity Suppose that $\leq$ is a well-ordering relation on A. For eac $B \in \mathfrak{P}(A) - \{\varnothing\}$ let $f(B)$ be the least element of B with respect to $\leq$. is clear that this defines a choice function for $\mathfrak{P}(A)$.

Sufficiency If f is a choice function for $\mathfrak{P}(A)$, then by the general princip of transfinite recursion and Hartogs' theorem there exist a unique ordin β and mapping $g : \boldsymbol{\beta} \to A$ such that $g(\alpha) = f(A - g[\boldsymbol{\alpha}])$ for all $\alpha < \beta$ an $g[\boldsymbol{\beta}] = A$; this mapping is bijective and therefore gives rise to a we ordering on A. [

The results of this section make it clear that there is a close relationshi between the following three questions:

Does every collection have a choice function?

Is every collection well-orderable?

Does every inductively ordered collection have a maximal element?

In fact none of these questions is answerable in basic set theory. We shall go further into the consequences of this in §7.3. In the meantime here are some particular cases in which these questions may be answered affirmatively.

Proposition 7.1.6

(a) *Every finite collection has a choice function* (The principle of finite choice).
(b) *Every countable collection is well-orderable.*
(c) *Every countable inductively ordered collection has a maximal element.*
(d) *If $\mathscr{B}$ is a collection of finite character such that $\bigcup \mathscr{B}$ is countable and if $A \in \mathscr{B}$, then $\mathscr{B}$ has a maximal element with respect to inclusion containing A.*

Proof

(a) If $\mathscr{A}$ is finite, then choice($\mathscr{A}$) is finite and non-empty, and hence has a maximal element [theorem 4.4.1] which must be a choice function for $\mathscr{A}$ [lemma 7.1.4(b)].
(b) Trivial.
(c) Immediate [(b) and theorem 7.1.2].
(d) If $\bigcup \mathscr{B} = \varnothing$, then $\mathscr{B} = \{\varnothing\}$ and the result is trivial. If not, then there exists a sequence (b_n) whose range is $\bigcup \mathscr{B}$. Now let $A_0 = A$. Once A_n has been defined, let A_{n+1} be the intersection of the elements of $\mathscr{A}$ containing $A_n \cup \{b_n\}$ if there are any; otherwise let $A_{n+1} = A_n$. In this way we recursively define an increasing sequence (A_n) in $\mathscr{B}$ whose range $\{A_n : n \in \omega\}$ therefore has a supremum B since $\mathscr{B}$ is inductively ordered by inclusion [proposition 7.1.3]. B is evidently a maximal element of $\mathscr{B}$ containing A. □

7.2 The axiom of countable dependent choice

We have already seen examples in §5.5 which illustrate the usefulness in mathematics of the axiom of countable choice. This axiom permits us, if we are given a sequence (A_n) of non-empty sets *in advance*, to choose an element x_n from each one. However, there are occasions when it is necessary to make a sequence of choices each of which is dependent on the previous one.

Example Let us try to prove that a partially ordered set is partially well ordered iff it contains no strictly decreasing sequences. Certainly one direction of this equivalence is straightforward: the image of a strictly decreasing sequence is a non-empty set with no minimal element. To prove the reverse implication we suppose that A is not partially well ordered, so that it has a non-empty subset B without a minimal element. Now choose an element x_0 in B and define a sequence (x_n) in B as follows: once x_n has been chosen, let x_{n+1} be any element of B less than x_n.

(Such an element exists because B has no minimal element.) The sequence (x_n) i clearly strictly decreasing.

This argument is undoubtedly plausible. However, it cannot be justifie by appeal to the axiom of countable choice because (using the tempora metaphor again) the choices involved are not simultaneous: x_{n+1} canno be chosen until the value of x_n is known. The necessity for making choice in this way arises sufficiently often for it to be worthwhile to single ou the set-theoretic principle which licenses the procedure.

Definition

The *axiom of countable dependent choice* (sometimes known as DC_ω fo short) asserts that if (A, r) is a small structure such that $\mathrm{dom}[r] = A$ an if $a \in A$ then there exists a sequence (x_n) in A such that $x_0 = a$ and $x_n \; r \; x_{n+}$ for all $n \in \boldsymbol{\omega}$.

We shall show now that the axiom of countable dependent choice is, a we should expect from the informal account we have given of it, at leas as strong as the axiom of countable choice.

Lemma 7.2.1

These three assertions are equivalent:

(i) *The axiom of countable choice.*

(ii) *Every countable set has a choice function.*

(iii) *If (A_n) is a small sequence of* disjoint *non-empty sets, then there exist a sequence (x_n) such that $x_n \in A_n$ for all $n \in \boldsymbol{\omega}$.*

Proof Exercise. □

Proposition 7.2.2

The axiom of countable dependent choice implies the axiom of countabl choice.

Proof Suppose that (A_n) is a small sequence of disjoint non-empty sets. Choos an element $a \in A_0$ and define a relation r on $\bigcup_{n \in \boldsymbol{\omega}} A_n$ by letting $x \; r \; y$ i there exists $n \in \boldsymbol{\omega}$ such that $x \in A_n$ and $y \in A_{n+1}$. Now clearly $\mathrm{dom}[r] =$ $\bigcup_{n \in \boldsymbol{\omega}} A_n$ and so by the axiom of countable dependent choice there exist a sequence (x_n) such that $x_0 \in A_0$ and $x_n \; r \; x_{n+1}$ for all $n \in \boldsymbol{\omega}$. It follow easily by induction that $x_n \in A_n$ for all $n \in \boldsymbol{\omega}$. This is enough to prove th axiom of countable choice [lemma 7.2.1]. □

Theorem 7.2.3

These three assertions are equivalent:

(i) *The axiom of countable dependent choice.*

(ii) *If (B, r) is a small structure such that* $\mathrm{dom}[r] = B$ *and B is non-empty then there exists a sequence* (y_n) *in B such that* $y_n\ r\ y_{n+1}$ *for all* $n \in \omega$ *(the value of* y_0 *is not stipulated).*

(iii) *Every partially ordered set which does not contain the image of a strictly decreasing sequence is partially well ordered.*

(i) ⇒ (ii) Trivial.

(ii) ⇒ (iii) Suppose that $(A, \leq)$ is not partially well ordered. So it has a non-empty subset B without a minimal element. Now $\mathrm{dom}[>_B] = B$ (since otherwise B would have a minimal element) and hence by hypothesis there exists a sequence (y_n) in B such that $y_n >_B y_{n+1}$ for all $n \in \omega$. Thus A contains the image of the strictly decreasing sequence (y_n).

(iii) ⇒ (ii) Suppose that (B, r) is a small structure such that $\mathrm{dom}[r] = B$. Let A be the set of all strings ϕ in String(B) such that

$$\phi(0)\ r\ \phi(1)\ r \ldots r\ \phi(n-1)$$

(where n is the length of the string ϕ). If we regard A as being partially ordered by the *opposite* of inclusion then it is evidently not partially well ordered since $\mathrm{dom}[r] = B$. Hence by hypothesis there exists a strictly decreasing sequence (ϕ_n) in A. Now $\{\phi_n : n \in \omega\}$ is a chain, and so if $\phi = \bigcup_{n \in \omega} \phi_n$, then ϕ is a sequence in B [lemma 2.2.1] such that $\phi(n)\ r\ \phi(n+1)$ for all $n \in \omega$. This is what we wanted.

(ii) ⇒ (i) Suppose that (A, r) is a structure such that $\mathrm{dom}[r] = A$ and $a \in A$. Let $B = \mathrm{Cl}_r(a)$. Then $\mathrm{dom}[r_B] = B$ since B is r-closed. Consequently by hypothesis there exists a sequence (y_n) in B such that $y_n\ r\ y_{n+1}$ for all $n \in \omega$. Now $B = \bigcup_{m \in \omega} r^m[a]$ [propositions 2.3.3(a) and 4.3.7(b)] and so $y_0 \in r^m[a]$ for some $m \in \omega$, i.e. there exist $z_0, z_1, \ldots, z_m \in B$ such that $z_0 = a$, $z_m = y_0$, and $z_n\ r\ z_{n+1}$ for $n < m$. If we let

$$x_n = \begin{cases} z_n & \text{for } n < m \\ y_{n-m} & \text{for } n \geq m, \end{cases}$$

then it is clear that $x_0 = a$ and $x_n\ r\ x_{n+1}$ for all $n \in \omega$. □

The axiom of countable dependent choice This axiom, which was first stated explicitly by Bernays in 1942, is widely used in analysis. Although it cannot be proved from the axiom of countable choice (Mostowski 1948) even in the presence of the axiom of purity (Jensen 1966), it seems to be just as plausible as countable choice: the motivation we gave for countable choice in §5.5 applies equally to countable dependent choice. Indeed dependent choice is probably the strongest choice axiom for which this sort of motivation can be given. The naturalness of dependent choice as an axiom has been emphasized fairly recently by Blair's demonstration in 1977 that it is equivalent to the Baire category theorem.

Exercises

1. Assuming the axion of countable dependent choice, show that if $(A, \leq)$ is an infinite partially ordered set, then A has either an infinite totally ordered subset or an infinite totally unordered subset. [Suppose that all totally unordered subsets of A are finite: show that every infinite subset B of A has a maximal totally unordered subset and therefore has an element b which is comparable with infinitely many elements of B.]
2. Show that the axiom of countable dependent choice is equivalent to the assertion that if $(A, \leq)$ is a partially ordered set and (D_n) is a sequence of cofinal subsets of A then there is an increasing sequence (x_n) in A such that $\{x_n : n \in \omega\}$ intersects every D_n.
3. Prove lemma 7.2.1.

7.3 The axiom of choice

The quest to remove unnecessary countability assumptions (essentially begun by Cantor) was one of the dominant leitmotivs of mathematics in the first half of this century, just as the quest to reintroduce them (because of the needs of computer science) has been dominant in the second half. The notion that there is nothing special about countability, which is most pervasive in the works of Bourbaki, makes assuming only that every *countable* set has a choice function seem a mite perverse.

Definition

The *axiom of local choice* (or just the *axiom of choice*) is the assertion that *every* set has a choice function.

The axiom of choice obviously implies the axiom of countable choice. Less obviously:

Proposition 7.3.1

The axiom of choice implies the axiom of countable dependent choice.

Proof Suppose that a, A, and r satisfy the hypotheses of the axiom of countable dependent choice. Then $r[x] \neq \varnothing$ for all $x \in A$. So by the axiom of choice there exists a mapping $f : A \to A$ such that for each $x \in A$ we have $f(x) \in r[x]$, i.e. $x \; r \; f(x)$. Now let $x_0 = a$ and once x_n has been defined let $x_{n+1} = f(x_n)$. Evidently $x_n \; r \; x_{n+1}$ for all $n \in \omega$. □

Remark The axiom of choice cannot be proved even if we assume countable dependent choice (Mostowski 1948) and purity (Feferman 1965).

The axiom of choice is by far the most important of the choice axioms not least because of the number of significant assertions in apparently unconnected parts of mathematics which can be shown to be equivalent to it. There are several forms which the axiom of choice commonly takes when it appears in proofs:

Theorem 7.3.2

These six assertions are equivalent:

(*i*) *The axiom of choice.*

(*ii*) *For every small family* $(A_i)_{i \in I}$ *of non-empty* [*disjoint*] *sets* $\Pi_{i \in I} A_i$ *is non-empty* (Multiplicative axiom).

(*iii*) *Every set is well-orderable* (Well-ordering principle).

(*iv*) *Every inductively ordered set has a maximal element* (Maximal principle).

(*v*) *If* $\mathscr{B}$ *is a set of finite character and* $A \in \mathscr{B}$, *then* $\mathscr{B}$ *has a maximal element with respect to inclusion containing* A (Teichmüller 1939; Tukey 1940).

(*vi*) $(\mathbf{Cn}, \leq)$ *is totally ordered.*

(i) ⇒ (ii) By hypothesis the set $\{A_i : i \in I\}$ has a choice function f: now $f(A_i) \in A_i$ for all $i \in I$ and so the family $(f(A_i))_{i \in I}$ is an element of $\Pi_{i \in I} A_i$.
(ii) ⇒ (i) If $\mathscr{A}$ is a set then $(A \times \{A\})_{A \in \mathscr{A} - \{\varnothing\}}$ is a small family of disjoint non-empty sets: the product set $\Pi_{A \in \mathscr{A} - \{\varnothing\}}(A \times \{A\})$ is therefore non-empty by hypothesis. But if $f \in \Pi_{A \in \mathscr{A} - \{\varnothing\}} A$ then $\mathrm{dom}(f(A)) \in A$ for all $A \in \mathscr{A} - \{\varnothing\}$ and so $A \mapsto \mathrm{dom}(f(A))$ $(A \in \mathscr{A} - \{\varnothing\})$ is a choice function for $\mathscr{A}$.
(i) ⇔ (iii) Zermelo's theorem.
(iii) ⇒ (iv) Theorem 7.1.2.
(iv) ⇒ (v) Suppose that $\mathscr{B}$ is a set of finite character and $A \in \mathscr{B}$. Then $\mathscr{B}$ is inductively ordered by inclusion [proposition 7.1.3], hence so also is $\mathscr{A} = \{B \in \mathscr{B} : A \subseteq B\}$, which therefore by hypothesis has a maximal element.
(v) ⇒ (i) Suppose that A is a set. Then choice(A) is of finite character [lemma 7.1.4(a)] and therefore by hypothesis has a maximal element f with respect to inclusion, which must be a choice function for A [lemma 7.1.4(b)].
(iii) ⇒ (vi) $(\mathbf{Wo}, \leq)$ is totally ordered [proposition 6.6.1]. If every set is well-orderable, then $\mathbf{Cn} = \mathbf{Wo}$ and so $(\mathbf{Cn}, \leq)$ is totally ordered.
(vi) ⇒ (iii) **Wo** is an unbounded initial subclass of **Cn** [proposition 6.7.7]. So if **Cn** is totally ordered, then $\mathbf{Wo} = \mathbf{Cn}$, and hence every set is well-orderable. □

The axiom of choice The first essential use of the axiom of choice seems to have been by Cantor in 1887–8: 'If M and N are two sets, then we understand by $M \,.\, N$ some third set which comes out of N by putting *in the place of each* individual element of N a set equivalent to the set M. It is now very easy to prove that *all* sets $M.N$ obtained by the method just described are equivalent to one another. The definition of the product of two cardinal numbers is based on this.' Cantor used the technique of making assertions which depend on the axiom of choice and claiming that they are 'very easy to prove' on several occasions. Indeed there is no evidence to suggest that Cantor ever doubted for a moment the validity of making arbitrarily many choices, a principle which, in Zermelo's words, he 'unconsciously and instinctively used everywhere and expressly stated nowhere' (Cantor 1932, p. 451).

We referred in §5.5 to the way in which implicit uses of the axiom of countable choice became common in the last quarter of the nineteenth century. However, apart from its use by Cantor the unrestricted axiom of choice does not seem to have been much used until after the appearance of Cantor's publications (1895, 1897*a*). Subsequently the partition principle (which depends on the axiom of choice but is weaker than it) was used in work on cardinal arithmetic in Germany by Cantor's pupil Felix Bernstein (unpublished thesis, 1901) (who was criticized for this by Levi (1902)) and in Italy by Burali-Forti (1896) (despite the antagonism of the members of Peano's circle, including Burali-Forti, to the axiom of countable choice). The axiom was also used implicitly by the Trinity mathematicians Whitehead (1902) and Hardy (1904). Russell came rather close to an explicit statement of the axiom in the work he contributed to Whitehead's paper (1902) when he stated as a postulate the assertion that every non-finite set is a disjoint union of countably infinite sets, which is equivalent to the axiom of choice. But it was only later that he came to see the implicit assumption Whitehead had made in his proof (in the same paper) that any family of cardinal numbers has a product; in his attempt to prove this assumption he was led to formulate the 'multiplicative axiom' explicitly in 1904.

However, Russell did not publish the multiplicative axiom until 1906; the first published formulation arose from a somewhat different line of research. In 1883 Cantor stated the well-ordering principle and, although he did not exploit it at that time, he evidently regarded it as an axiom, describing it as 'a law of thought which appears to me to be fundamental, rich in consequences, and particularly remarkable for its general validity' (1883*a*, p. 550). However, in later publications (Cantor 1895, 1897*a*) he had retreated from this position and regarded the well-ordering principle as a conjecture requiring proof: he appears to have attempted such a proof in 1896 and communicated it to Hilbert (see letter to Jourdain, 1903). If the 'proof' he gave to Hilbert was the one he later sent to Dedekind (letter 1899*b*), then it is not convincing. However, in his 1900 lecture to the International Congress of Mathematicians in Paris Hilbert referred to the well-ordering principle as a 'theorem of Cantor', but posed the problem of finding a definite well-ordering on the real line. Zermelo (a colleague of Hilbert at Göttingen) formulated the axiom of choice (*Auswahlaxiom*) explicitly and deduced the well-ordering principle from it in 1904; the more demanding problem of finding a *definite* well-ordering on **R** has been shown not to be soluble even if we assume the axiom of choice (Feferman 1965). (Put more informally, this means that the axiom of choice guarantees the existence of a well-ordering on **R** without enabling us to define one.)

Maximal principles Hausdorff (1909, p. 301) asserted that (in modern terminology) if $\mathscr{A}$ is a set of subsets of a fixed set A and if the union of every well ordered chain in $\mathscr{A}$ is an element of $\mathscr{A}$ then $\mathscr{A}$ has a maximal element with respect to inclusion. However, Hausdorff's widely read textbook on *Mengenlehre* (1914) does not mention this result (although it does contain the closely related result of exercise 2 below, which is often referred to as 'Hausdorff's maximality principle') and neither he nor anyone else seems to have appreciated its usefulness then. (The result does appear in the 1927 second edition of *Mengenlehre* though.)

When Kuratowski rediscovered the maximal principle in 1922, it was quite explicitly as part of a reductionist programme stemming from Zermelo's proofs of the well-ordering principle. Zermelo's first proof (1904) made use of transfinite induction on the ordinals and was therefore haunted by the spectre of the Burali-Forti paradox in the eyes of some. For example, Hobson (1905, p. 185): 'The non-recognition of the existence of "inconsistent" aggregates, which existence, on the assumption of Cantor's theory cannot be denied, introduces an additional element of doubt as regards this proof' (cf. Hardy 1906, p. 17n.). In order to answer this criticism Zermelo published another proof in 1908 which eliminated the previous use of ordinals by means of an extension of Dedekind's *Kettentheorie*. The form of Zermelo's axiomatization of set theory (1908*a*) seems to have been strongly influenced by this second proof: crudely speaking, he chose the weakest natural-seeming axioms which would justify his proof. As a result his system was not strong enough to contain ordinal arithmetic without an additional postulate on the existence of transfinite numbers. But in the following years numerous important results were proved from the axiom of choice by transfinite induction. (For example, Steinitz (1910) demonstrated by this means the existence and uniqueness up to isomorphism of the algebraic closure of an arbitrary field.) Kuratowski turned Zermelo's method for eliminating ordinals into a general procedure and hence obtained the maximal principle: the result was a method by which proofs which use transfinite induction could be transformed into proofs which do not and which could therefore be formalized in Zermelo's system. (Whitehead and Russell came rather close to stating the maximal principle in their presentation of Zermelo's theorem in *Principia Mathematica* (1910–13, *258).)

After Kuratowski the maximal principle was used occasionally but did not achieve widespread fame because the advantages he claimed for it were axiomatic and aesthetic rather than practical: mathematicians have always been loath to give up convenient tools for the sake of logical purity. The second rediscovery of the maximal principle by Zorn (1935) was decisive because he gave convincing evidence of its *usefulness* and because he had recently emigrated from Hamburg to New England, where the result soon became widely known as 'Zorn's lemma' (the name it has kept ever since) among the active research communities of Yale and Princeton (see Campbell 1978).

The final step in the popularization of the maximal principle was taken by Bourbaki (1939), who included the version for abstract partial order relations—first stated by Bochner (1928)—and the Teichmüller/Tukey principle, and exploited them frequently in the subsequent parts of his treatise. Bourbaki's presentation was the fulfilment of Kuratowski's reductionist intentions, to the extent that ordinals are never defined in the text of his work, but only in the exercises. However, Bourbaki's reasons were not foundational, as Kuratowski's had largely been, since Bourbaki's system is quite strong enough to allow the definition of the ordinals; rather they were aesthetic, based on Bourbaki's preference for what he saw as purely algebraic methods.

Cardinal arithmetic and the axiom of choice The comparability of cardinals was conjectured by Cantor (1895, 1897*a*). Peirce seems to have deduced it from the axiom of choice in a lecture of 1897 which was not published until 1933 and consequently had no lasting influence. Of course it is clear that cardinal comparability follows from the well-ordering principle but the converse was not proved until 1915 by Hartogs.

If we assume the axiom of choice then $\mathbf{Wo} = \mathbf{Cn}$, from which it follows that the results we proved in §5.6 about well-orderable small cardinals are true for all small cardinals. This brings about a considerable simplification in cardinal arithmetic, since addition and multiplication of small infinite cardinals become trivial: the axiom of choice implies that $\mathfrak{a} + \mathfrak{b} = \mathfrak{a}\mathfrak{b} = \max(\mathfrak{a}, \mathfrak{b})$ for all small infinite cardinals. Indeed the latter assertion, like many others in cardinal arithmetic, is actually equivalent to the axiom of choice (Tarski 1924). So if we assume the axiom of choice, only cardinal exponentiation remains non-trivial.

Is the axiom of choice true? Zermelo's derivation of the well-ordering principle from the axiom of choice in 1904 and a purported proof of the contrary by König (1905) were much discussed in *Mathematische Annalen* (Borel 1905; Jourdain 1905; Bernstein 1905*b*; Schönflies 1905), *Bulletin de la Société Mathématique de France* (Baire *et al.* 1905), *Proceedings of the London Mathematical Society* (Hobson 1905; Hardy 1906; Dixon 1906

Jourdain 1906; Russell 1906) and elsewhere. Since then many mathematicians have doubted the axiom of choice. For instance, Littlewood (1926): 'Reflection makes the intuition of its truth doubtful, analysing it into prejudices derived from the finite case, and short of intuition there seems no evidence in its favour.' What is clear is that the sort of temporal motivation which we gave for the axiom of countable dependent choice is not available in the general case. A quasi-temporal argument can perhaps be given in favour of the *axiom of well ordered choice* (first proposed by Hardy (1906)), which asserts that the image of every transfinite sequence $(A_\alpha)_{\alpha<\beta}$ of non-empty sets has a choice function, but even here the temporal analogy seems rather far-fetched when β is uncountable.

The axiom of choice has many elegant consequences, but that is an argument for its mathematical interest, not for its truth. Moreover, arguments which make essential use of the axiom occur only in the most abstract parts of mathematics. This makes the question of its truth one to which many mathematicians—in particular, applied mathematicians—may happily remain indifferent. Furthermore, there is no possibility of the axiom being either confirmed or refuted in any physical interpretation: there simply do not exist theorems of the form 'If the Forth Bridge blows down in a gale then the axiom of choice is true.'

In number theory the axiom of choice is hardly usable at all, and even in geometry its consequences are not generally assertions which our intuition gives us grounds for either accepting or rejecting independent of our views of the axiom itself. One exception to this may be the theorem of Banach and Tarski (1924) which asserts that any sphere in $\mathbf{R}^3$ of unit radius may be partitioned into a finite number of (non-measurable) pieces which can be rotated in $\mathbf{R}^3$ to form a partition of *two* disjoint spheres of unit radius. Many people find this result implausible. However, the significance of this should not be overstated. It may be interpreted rather as showing that our intuitions about 'geometrical' sets in Euclidean space are represented mathematically by theorems about *measurable* sets. (The existence of non-measurable sets depends on the axiom of choice.)

The most decisive way of justifying the axiom of choice would be to convince oneself of the truth of the AXIOM OF STRONG CHOICE, of which it is a trivial consequence. Since the AXIOM OF STRONG CHOICE is purely logical, its validity would in no way be dependent upon the particular features of one's conception of set theory (although it might, of course, depend upon a view about the logical status of mathematical objects in general). The formal consistency of making this assumption can hardly be doubted, but it ascribes to us abilities which I for one am not aware of possessing. (Perhaps this is the point at which bringing in the notion of God would be helpful.)

A free-swinging realist who is unhappy about claiming the full strength of the AXIOM OF STRONG CHOICE can nevertheless justify the axiom of choice by recalling the maximal conception of the construction of power collections which we alluded to in the note on 'Predicative set theory' in §1.5. According to this account the axiom of choice is trivially true since if on each day we create *all* subcollections of the previous day we will in particular create a subcollection containing one element from each member of the previous day. Most pure mathematicians nowadays seem to be realists of this sort.

Indeed it may now seem quixotic to adopt any system of set theory intermediate in strength between a totally predicative one and the strongly realist one with the axiom of choice. But this is wrong: if we believe, as realists do, that mathematical objects exist independent of our ability to describe them, we may nevertheless think it is opaque what sort of object an indefinable collection is and where our intuitions about it derive from. On this view it is dangerous to claim the existence of an object one cannot describe: 'Whereof one cannot speak thereof one must be silent', to hijack a slogan (Wittgenstein 1922, pp. 27 and 189).

Example The set $\mathscr{A}$ of partial order relations on a set A is of finite character, and the maximal elements of $\mathscr{A}$ are precisely the total order relations on A (see exercise 4 of §2.6). It therefore follows from the Teichmüller/Tukey property that every partial order relation on A is contained in a total order relation (Szpilrajn 1930). This result also follows from the well-ordering property by the method described in example 4 of §6.2.

Exercises

1. Show that the axiom of choice is equivalent to the assertion that every partially ordered set has a maximal totally ordered subset. [*Necessity* Use the Teichmüller/Tukey property. *Sufficiency* Prove the maximal principle.]
2. Assuming the axiom of choice, prove that every partially ordered set $(A, \leq)$ has a maximal totally unordered subset. [*First method* Use the well-ordering principle. *Second method* Show that the set of all totally unordered subsets of A is of finite character and then use the Teichmüller/ Tukey property.]
3. Show that the axiom of choice is equivalent to the assertion that if A, B are sets then every surjective mapping $A \to B$ has a right inverse. (Compare exercise 1 of §2.2.)
4. Assuming the axiom of choice, prove that a relation is a partial ordering [respectively partial well-ordering] on the set A iff it is the intersection of a collection of total orderings [respectively well-orderings] on A.
5. Assuming the axiom of choice, prove that every partially ordered set $(A, \leq)$ has a cofinal partially well ordered subset. [Let $\mathscr{A}$ be the set of partially well ordered subsets of A. Apply the maximal property to $\mathscr{A}$ with the partial ordering 'is an initial subset of'.]
6. (a) Assuming the axiom of choice, show that if $(A_i)_{i \in I}$ and $(B_i)_{i \in I}$ are small families such that $2 \leq \mathrm{card}(A_i) \leq \mathrm{card}(B_i)$ for all $i \in I$, then $\mathrm{card}(\dot{\bigcup}_{i \in I} A_i) \leq \mathrm{card}(\prod_{i \in I} B_i)$.
   (b) Show that the axiom of choice is equivalent to the assertion that if $(A_i)_{i \in I}$ and $(B_i)_{i \in I}$ are small families such that $\mathrm{card}(A_i) < \mathrm{card}(B_i)$ for all $i \in I$, then $\mathrm{card}(\dot{\bigcup}_{i \in I} A_i) < \mathrm{card}(\prod_{i \in I} B_i)$ (*König's inequality*). [*Necessity* Show

first that for any mapping $f : \dot{\bigcup}_{i \in I} A_i \to \prod_{i \in I} B_i$ there is a family $(b_i)_{i \in I}$ in $\prod_{i \in I} B_i$ such that $b_i \notin \mathrm{pr}_i[f[A_i \times \{i\}]]$ for all $i \in I$. *Sufficiency* $0 < 1$.]

7. Deduce Cantor's theorem from König's inequality. [*Hint* $1 < 2$.]
8. (a) Assuming the axiom of choice, show that $2^{\aleph_0} \neq \aleph_\omega$. [Apply König's inequality to a sequence (A_n) such that $\mathrm{card}(A_n) = \aleph_n$ for all $n \in \boldsymbol{\omega}$ and a set B such that $\mathrm{card}(B) = \aleph_\omega$.]
   (b) Show that the axiom of choice is not necessary in (a). [Observe that if $2^{\aleph_0} = \aleph_\omega$ then in particular $2^{\aleph_0}$ is well-orderable.]
9. Prove that the axiom of choice is equivalent to the assertion that every small tree has a branch. [*Sufficiency* If A is a set, show that the set $\mathscr{A}$ of all graphs of injective mappings from initial segments of **On** to A is a tree with respect to inclusion and deduce that A is well-orderable.]
10. A *graph* is a structure (A, r) such that r is a symmetric relation. If $n \in \boldsymbol{\omega}$ then an *n-colouring* of a subcollection B of A is a mapping $f : B \to \boldsymbol{n}$ such that $x\, r\, y \Rightarrow f(x) \neq f(y)$.
    (a) Assuming the axiom of choice, prove that the set $\mathscr{A}_n$ of n-colourable subsets of a graph A is of finite character (*Colouring theorem*). [Apply the maximal principle to the set of n-colourings f of subsets B of A such that for every finite $F \subseteq A$ f can be extended to an n-colouring of $B \cup F$.]
    (b) Deduce that if we assume the axiom of choice then A has a maximal n-colourable subcollection.
11. Show that the axiom of choice is equivalent to the assertion that if $(A_i)_{i \in I}$ is a family of non-empty *classes* indexed by a set I then $\prod_{i \in I} A_i$ is non-empty.
12. Deduce the well-ordering property directly from the maximal property. [Use exercise 4 of §7.1.]

7.4 The axiom of global choice*

Definition

The *axiom of global choice* asserts that there is a definite choice function on **D**.

The axiom of global choice is a particular case of the AXIOM OF STRONG CHOICE. To see this, observe that for each $A \in \mathbf{D} - \{\varnothing\}$ it is the case that $(\exists x)(x \in A)$ and so (assuming the AXIOM OF STRONG CHOICE) $(\exists! x)(x \in A)$, i.e. $\iota x(x \in A)$ exists: hence the function $(\iota x(x \in A))_{A \in \mathbf{D} - \{\varnothing\}}$ which this term defines is a definite choice function for **D**.

Proposition 7.4.1

The axiom of global choice implies the axiom of choice.

Proof Obvious. □

Proposition 7.4.2

These two assertions are equivalent:

(i) The axiom of global choice.

(ii) There exists a definite well ordering on **D**.

* This section may be omitted without loss of continuity.

(i) ⇒ (ii) It is enough [theorem 7.1.5] to show that $\mathfrak{P}(\mathbf{D})$ has a definite choice function. So define a function $f : \mathfrak{P}(\mathbf{D}) \to \mathbf{D}$ by $f(A) = \langle x : x \in A \rangle$. Now by hypothesis $\mathbf{D}$ has a definite choice function g. Hence $g \circ f$ is a definite choice function for $\mathfrak{P}(\mathbf{D})$ since $f(A) \subseteq A$ for all $A \in \mathfrak{P}(\mathbf{D})$.
(ii) ⇒ (i) For each $A \in \mathbf{D} - \{\varnothing\}$ let $f(A)$ denote the least element of A with respect to a definite well ordering on $\mathbf{D}$. □

7.5 The generalized continuum hypothesis

Of all the operations of basic set theory the formation of power sets is the most important but the least understood. In particular, the question of its richness is central to at least one way of viewing the axiom of choice: as we remarked in §7.3, if the power operation is (in some sense) maximal then we may expect the axiom of choice to be true. In this section we shall establish a result at the opposite extreme: if the power operation is minimal then the axiom of choice is true. The sense of minimality we intend is that the cardinality of a power set should never be larger than Cantor's theorem allows.

Definition

The *generalized continuum hypothesis* is the assertion that $2^{\mathfrak{a}}$ is a successor of $\mathfrak{a}$ for all $\mathfrak{a} \in \mathbf{Cn} - \boldsymbol{\omega}$.

Let us recall (§6.7) that if $\mathfrak{a} \in \mathbf{Cn}$ then there is a unique $\mathfrak{a}^+ \in \mathbf{Wo}$ with the following properties:

$$\mathfrak{a}^+ \nleq \mathfrak{a}; \quad \mathfrak{a}^+ \leq 2^{2^{\mathfrak{a}^2}}; \quad \mathfrak{b} < \mathfrak{a}^+ \Rightarrow \mathfrak{b} \leq \mathfrak{a}.$$

Lemma 7.5.1

If $\mathfrak{a} \in \mathbf{Cn}$ *and* $\mathfrak{a} + \mathfrak{a}^+ = \mathfrak{a}\mathfrak{a}^+$ *then* $\mathfrak{a} \in \mathbf{Wo}$.

Proof Let A and B be sets such that $\operatorname{card}(A) = \mathfrak{a}$ and $\operatorname{card}(B) = \mathfrak{a}^+$. By hypothesis there exist disjoint sets A' and B' such that $A \times B = A' \cup B'$. Suppose first that $(\exists a \in A)(\forall y \in B)((a, y) \in A')$. Then there is an injective mapping $B \to A'$ given by $y \mapsto (a, y)$. So $\mathfrak{a}^+ \leq \mathfrak{a}$. Contradiction. Hence, since B is well-orderable, we can define a mapping $f : A \to B$ such that $(x, f(x)) \in B'$ for all $x \in A$. The mapping $A \to B'$ given by $x \mapsto (x, f(x))$ is evidently injective and therefore $\mathfrak{a} \leq \mathfrak{a}^+$. Consequently $\mathfrak{a}$ is well-orderable since $\mathfrak{a}^+$ is. □

Theorem 7.5.2

The generalized continuum hypothesis implies the axiom of choice.

Proof Suppose not. So the generalized continuum hypothesis holds but there is a cardinal $\mathfrak{a} \in \mathbf{Cn} - \mathbf{Wo}$. Let $\mathfrak{b} = 2^{\mathfrak{a}+\aleph_0}$. Then $\mathfrak{a} < \mathfrak{b}$ and so $\mathfrak{b}$ is not well-orderable. Also $\mathfrak{b} \le 2\mathfrak{b} = 2^{\mathfrak{a}+\aleph_0+1} = 2^{\mathfrak{a}+\aleph_0} = \mathfrak{b}$, so that $2\mathfrak{b} = \mathfrak{b}$. Hence $\mathfrak{b}^2 \le (2^{\mathfrak{b}})^2 = 2^{2\mathfrak{b}} = 2^{\mathfrak{b}}$ and therefore

$$\mathfrak{b}^+ \le 2^{2^{\mathfrak{b}^2}} \le 2^{2^{2^{\mathfrak{b}}}}. \tag{7.5.1}$$

We intend to show now that $(2^{\mathfrak{b}})^+ = \mathfrak{b}^+$. For if not then $\mathfrak{b}^+ < (2^{\mathfrak{b}})^+$ and so $\mathfrak{b}^+ \le 2^{\mathfrak{b}}$. Now $\mathfrak{b} \ge \aleph_0$ and so $\mathfrak{b} + 1 = \mathfrak{b}$. Hence $\mathfrak{b} \le \mathfrak{b} + \mathfrak{b}^+ \le 2^{\mathfrak{b}+1} = 2^{\mathfrak{b}}$. But if $\mathfrak{b} + \mathfrak{b}^+ = \mathfrak{b}$ then $\mathfrak{b}^+ \le \mathfrak{b}$, which is absurd. Hence $\mathfrak{b} + \mathfrak{b}^+ = 2^{\mathfrak{b}}$ [generalized continuum hypothesis]. So $2^{\mathfrak{b}} = \mathfrak{b} + \mathfrak{b}^+ \le \mathfrak{b}\mathfrak{b}^+ \le (2^{\mathfrak{b}})^2 = 2^{\mathfrak{b}}$ and therefore $\mathfrak{b} + \mathfrak{b}^+ = \mathfrak{b}\mathfrak{b}^+$. Consequently $\mathfrak{b}$ is well-orderable. Contradiction. So $\mathfrak{b}^+ = (2^{\mathfrak{b}})^+$. Similar arguments show that $(2^{\mathfrak{b}})^+ = (2^{2^{\mathfrak{b}}})^+$ and that $(2^{2^{\mathfrak{b}}})^+ = (2^{2^{2^{\mathfrak{b}}}})^+$. It follows that $\mathfrak{b}^+ = (2^{2^{2^{\mathfrak{b}}}})^+ \not\le 2^{2^{2^{\mathfrak{b}}}}$, contradicting (7.5.1). □

Remark Assume the axiom of choice: then $\mathfrak{a}^+ \le 2^{\mathfrak{a}}$ for every aleph $\mathfrak{a}$; and the generalized continuum hypothesis is equivalent to the assertion that $\mathfrak{a}^+ = 2^{\mathfrak{a}}$ for every aleph $\mathfrak{a}$. (This equivalence also holds, less obviously, in the presence of the axiom of purity (Rubin 1960).)

The continuum hypothesis The conjecture that $2^{\aleph_0} = \aleph_1$ was made by Cantor (1878); it got the name 'continuum hypothesis' from Bernstein (thesis, 1901). The generalized form appeared in Hausdorff (1908). Cantor devoted a great deal of time to investigating the continuum question and on several occasions believed briefly that he had settled it. Hilbert (1900) regarded answering the continuum question as one of the most important challenges in mathematics. He sketched a purported proof of the continuum hypothesis in 1925 based on a classification of the elements of ${}^{\omega}\omega$ into orders of recursive definability, but the details were never completed and Zermelo is reported as saying that 'no-one understood what he meant' (Levy 1964, p. 89). In fact the continuum hypothesis is not provable in our system even if we assume the axioms of choice and purity (Cohen 1963–4).

The constructibility hypothesis At the opposite extreme to the maximal conception of the formation of power collections, which we have already mentioned, is the *minimal* conception according to which the only collections created at each stage are those forced on us by the axioms of **GA**. Fraenkel (1922*a*) suggested adding an axiom of restriction (*Axiom der Beschränktheit*) to achieve this, but did not formalize the notion. The first satisfactory formulation of such an axiom is due to Gödel (1938), who defined in the language of set theory a much more restrictive hierarchy consisting (roughly) of collections that can be defined by means of a formula which refers only to collections which have already been created. The *constructibility hypothesis* is the assertion that every set occurs in this hierarchy. Gödel showed that if set theory is consistent then it remains consistent if we add the constructibility hypothesis and the axiom of purity. The significance of this for us lies in the fact (Gödel 1938 again) that the generalized continuum hypothesis is provable in this restrictive theory. So if **GA** is consistent, then it remains consistent if we add the generalized continuum hypothesis (or *a fortiori* the axiom of choice).

At first sight the principle of ontological parsimony which encouraged us to eliminate un-well-founded collections makes the constructibility hypothesis—like the axiom of purity—an attractive assumption, since it asserts that every set occurs in a highly restrictive hierarchy in which only those collections essential to the theory are created. However, the image of the set-theoretic universe which this forces on us seems very implausible: it is difficult to find a reason for believing that the quasi-predicative creation process represented by the constructible hierarchy and the impredicative process represented by the traditional hierarchy should both result in the same collections being created. In any case, the constructibility hypothesis is of no help to us in deciding about the axiom of choice as long as we reject purity: if the collection of individuals is formless, no amount of care in limiting the construction of the hierarchy can change that.

Undecidable sentences and second-order axioms Goodstein's theorem is not decidable in first-order Peano arithmetic. The continuum hypothesis is not decidable in first-order set theory with the axiom of choice. Both are provable in certain stronger systems (e.g. set theory in the case of Goodstein's theorem, set theory with the constructibility hypothesis in the case of the continuum hypothesis). In both cases, too, it is difficult (but perhaps not impossible) to come to a view about the truth of the result independent of views about the truth of the axioms of the stronger system in which the result is known to be provable. So if most mathematicians believe Goodstein's theorem but not the continuum hypothesis—their names indicate this already—it can only be because most mathematicians believe the axioms of basic set theory but do not believe the constructibility hypothesis.

But there is another similarity: being shown the proofs that they are undecidable in the weaker system in question does not weaken one's conviction that they *are* decidable questions. This is in contrast to the position for the questions—are there any individuals? how many days are there?—on which we laid down axioms in Chapters 1 and 3. There was never any suggestion then of these being questions which had answers if we could only see them; they were matters to be settled by legislation.

The reason for this difference is that our failure to decide Goodstein's theorem in Peano arithmetic, or the continuum hypothesis in basic set theory, can be traced to our failure to express the (platonistic) imagined content of a second-order axiom by means of a first-order axiom scheme:

we are referring here to the induction axiom in Peano arithmetic and the axiom of subcollections in basic set theory.

When we first naively write down these axiom schemes we see them as second-order assertions. (Even this distinction is by no means a naive one.) We intend—or at any rate the set-theoretic platonist intends—$\mathfrak{P}(A)$ to contain *all* the subsets of A in as strong a sense of 'all' as possible; and in the same way the number-theoretic platonist intends induction to apply to *all* subsets of $\boldsymbol{\omega}$. But the first-order axiom schemes can only make these claims for sets which can be defined by means of formulae in the language of the first-order theory in question. There seems to be no way available to us of expressing the strong sense of 'all' which we need, other than by saying it louder, in the manner of Czech border guards.

The point has been made by Kreisel (1967); our feeling that these questions *ought* to be decidable stems from the fact that they are indeed decidable in the second-order systems. So much the worse for first-order logic, you might think. But Kreisel's remark is an explanation for an observed fact about our psychology, not a proposed programme for settling the continuum problem. It would be better to say: the continuum question is decidable in second-order set theory to the extent that second-order logic is decidable. The problems involved here are exactly the same as before: the reformulation in second-order terms transfers the difficulties from set theory to logic; it does not solve them.

Exercises

1. (Sierpinski 1924) Show that the continuum hypothesis is equivalent to the assertion that $\aleph_2^{\aleph_0} > \aleph_1^{\aleph_0}$.
2. Prove that the generalized continuum hypothesis is equivalent to the conjunction of the axiom of choice and the assertion that $2^{\mathfrak{a}} = \mathfrak{a}^+$ for every aleph $\mathfrak{a}$.

7.6 Further exercises

These exercises are intended to illustrate how the axioms we have been studying in this chapter are used in other parts of mathematics: in each one an elementary knowledge of the subject in question is assumed.

All the collections referred to in this section are assumed to be small.

Real analysis

1. Let us define an equivalence relation $\approx$ on $\mathbf{R}$ so that $x \approx y$ iff $x - y \in \mathbf{Q}$. Assume the axiom of choice.
   (a) Show that there is a subset V of $[0, 1]$ which has exactly one point in common with each equivalence class.
   (b) If $A \subseteq V$ is measurable, show that A is null (i.e. has measure zero).
   (c) Show that V is not measurable.

2. A real number a is said to be an *accumulation point* of a set $A \subseteq \mathbf{R}$ if for every $\varepsilon > 0$ there exists $x \in A$ such that $0 < |x - a| < \varepsilon$; a is said to be a *sequential accumulation point* of A if there is an injective sequence (x_n) in A which converges to a.
   (a) Show that every sequential accumulation point of a subset A of $\mathbf{R}$ is an accumulation point of A.
   (b) Prove the converse from the axiom of countable choice.
   (c) Prove that every bounded infinite subset of $\mathbf{R}$ has a sequential accumulation point.
   (d) Prove that every bounded non-finite subset of $\mathbf{R}$ has an accumulation point.
3. Suppose that $f: \mathbf{R} \to \mathbf{R}$ is a mapping. We say that f is *continuous* at a point $c \in \mathbf{R}$ if for every $\varepsilon > 0$ there exists $\delta > 0$ such that $|f(x) - f(c)| < \varepsilon$ whenever $|x - c| < \delta$; we say that f is *sequentially continuous* at c if for every sequence (x_n) converging in $\mathbf{R}$ to c the sequence $(f(x_n))$ converges to $f(c)$.
   (a) Show that if f is continuous at c then it is sequentially continuous at c.
   (b) (Cantor) Assuming the axiom of countable choice, show that if f is sequentially continuous at c then it is continuous at c.
   (c) (Sierpinski 1918) Without assuming the axiom of countable choice, show that if f is sequentially continuous at every point of $\mathbf{R}$ then it is continuous at every point of $\mathbf{R}$. [If $c \in \mathbf{R}$ and $\varepsilon > 0$, prove first that there exists $\delta > 0$ such that $|f(r) - f(c)| < \varepsilon$ for every *rational* number r such that $|r - c| < \delta$.]

Linear algebra

Suppose that V is a vector space over a field F.

(a) If $B \subseteq V$, prove that these three assertions are equivalent:
   (i) B is a basis for V;
   (ii) B is maximal among the linearly independent subsets of V;
   (iii) B is minimal among the spanning sets of V.

(b) Assuming the axiom of choice, prove that every linearly independent subset of V is contained in a basis for V. [*First method* Use the Teichmüller/Tukey property. *Second method* Index the elements of V by ordinals and define recursively an increasing transfinite sequence of linearly independent subsets of V.]

(c) Show without using the axiom of choice that if V has a finite spanning set then it has a basis.

(d) Show that if B and C are bases for V and either B or C is finite, then $\operatorname{card}(B) = \operatorname{card}(C)$.

(e) Assuming the axiom of choice, generalize (d) to the case when neither B nor C is finite. [If $x = \Sigma_{y \in C} \lambda_{xy}$ is the unique expression for x as a linear combination of elements of C, let $C_x = \{y \in C : \lambda_{xy} \neq 0\}$: show that C_x is finite and $C = \bigcup_{x \in B} C_x$.]

Topology

1. A topological space is said to be *separable* if it has a countable dense subset, and *second countable* if its topology has a countable base.
   (a) Show that every separable metric space is second countable.

(b) Assuming the axiom of coutable choice, show that every second countable topological space is separable.
(c) Deduce (assuming the axiom of countable choice) that every subspace of a separable metric space is separable.

2. A topological space X is said to be a *countable chain condition* (or ccc for short) space if every collection of pairwise disjoint open subsets of X is countable.
(a) Show that every separable space is ccc.
(b) Assuming the axiom of choice, prove that every ccc metric space is separable. [For each $n \in \omega$ show that there is a maximal subset A_n of X with the property that $d(x, y) \geq 1/n$ for all $x, y \in A_n$; then show that $\bigcup_{n \in \omega} A_n$ is a countable dense subset of X.]

3. A topological space is said to be *semiparacompact* if for every open covering $\mathscr{U}$ of X there exists a sequence $(\mathscr{V}_n)$ of collections of open subsets of X such that:
if $n \in \omega$ and $V \in \mathscr{V}_n$ then there exists $U \in \mathscr{U}$ such that $V \subseteq U$;
$\bigcup_{n \in \omega} \mathscr{V}_n$ is a covering of X;
if $x \in X$ then for each $n \in \omega$ there exists an open set $W \ni x$ which intersects only finitely many elements of $\mathscr{V}_n$.
(a) Show that every second countable topological space is semiparacompact.
(b) Deduce that every separable metric space is semiparacompact.
(c) Assuming the axiom of choice, prove that every metric space is semiparacompact. [Suppose that $\mathscr{U}$ is an open covering of the metric space X. Index $\mathscr{U}$ by ordinals so that (say) $\mathscr{U} = \{U_\beta : \beta < \alpha\}$. For each $n \in \omega$ and $\beta < \alpha$ let

$$S_{n,\beta} = \{x \in U_\beta : d(x, X - U_\beta) \geq 2^{-n}\}$$

$$T_{n,\beta} = S_{n,\beta} - \bigcup_{\gamma < \beta} S_{n+1,\gamma}$$

$$W_{n,\beta} = \{x \in X : d(x, T_{n\beta}) < 2^{-n-3}\}$$

$$\mathscr{V}_n = \{W_{n\beta} : \beta < \alpha\}.]$$

4. Suppose that X is a metric space. X is said to be *complete* if every decreasing sequence (A_n) of non-empty closed sets in X with $\operatorname{diam}(A_n) \to 0$ has a non-empty intersection. Prove the following results.
(a) If X is complete, then every Cauchy sequence in X converges.
(b) If X is separable and every Cauchy sequence in X converges, then X is complete.
(c) If we assume the axiom of countable choice, we can do without separability in (b).
(d) If X is compact, then it is complete.

5. (a) An *ultrametric* on X is a mapping $d : X \times X \to \mathbf{R}$ such that:

$$d(x, y) = d(y, x);$$

$$d(x, y) \geq 0;$$

$$d(x, y) = 0 \text{ iff } x = y;$$

$$d(x, z) \leq \max(d(x, y), d(y, z)).$$

Show that every ultrametric is a metric.

(b) For any two mappings $f, g : \omega \to A$ let $v(f,g)$ be the least $n \in \omega$ such that $f(n) \neq g(n)$ if it exists, and ∞ otherwise. We thus define a mapping $v : {}^{\omega}A \times {}^{\omega}A \to \omega^{+}$. Show that:

$$v(f,g) = v(g,f);$$

$$v(f,g) = 0 \text{ iff } f = g;$$

$$v(f,h) \geq \min(v(f,g), v(g,h)).$$

(c) Prove that if we let $d(f,g) = 2^{-v(f,g)}$ for all $f, g \in {}^{\omega}A$ (with the convention that $2^{-\infty} = 0$) then $({}^{\omega}A, d)$ is a complete ultrametric space.

6. (a) (Blair 1977; Goldblatt 1985) Prove that the axiom of countable dependent choice is equivalent to the assertion that if X is complete and (U_n) is a sequence of dense open subsets of X, then $\bigcap_{n \in \omega} U_n$ is dense in X (*Baire category theorem*). [*Necessity* If $\bigcap_{n \in \omega} U_n$ is not dense in X, it is disjoint from some non-empty open set V: use the axiom of countable dependent choice to define a sequence (V_n) of non-empty open sets in X such that $V_0 = V$, $\mathrm{Cl}(V_{n+1}) \subseteq V_n \cap U_n$ and $\mathrm{diam}(\mathrm{Cl}(V_n)) \to 0$. *Sufficiency* Suppose $(A, \leq)$ is a partially ordered set and (D_n) is a sequence of cofinal subsets of A. If $\mathscr{B}$ is the set of all increasing sequences in A, show that $\mathscr{B}$ is closed (and hence complete) in ${}^{\omega}A$ with the metric defined in exercise 5. Let $U_p = \{(x_n) \in \mathscr{B} : x_n \in D_p \text{ for some } n \in \omega\}$ and show that each U_p is a dense open subset of $\mathscr{B}$.]

(b) Assuming the axiom of countable dependent choice, deduce that if X is complete and $\{x\}$ is not open for any $x \in X$, then X is not countable.

Functional analysis

1. Suppose that E is a Hilbert space over $\mathbf{R}$ or $\mathbf{C}$ with inner product $(x, y) \mapsto (x|y)$.
   (a) Show that an orthonormal subset of E is total (*aka* complete) iff it is maximal among orthonormal subsets of E.
   (b) Assuming the axiom of choice, prove that E contains a total orthonormal set.

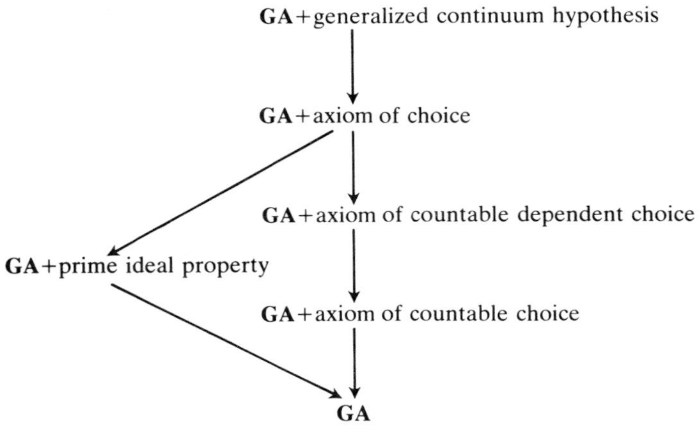

Figure 3

(c) If B and C are both total orthonormal subsets of E, show (assuming the axiom of choice) that $\operatorname{card}(B) = \operatorname{card}(C)$. [Deal first with the case when either B or C is finite. In the other case, let $C_x = \{y \in C : (x|y) \neq 0\}$: show that C_x is countable by using Bessel's inequality; then note that $C_x = \bigcup_{x \in B} C_x$.]

(d) If E is separable, prove the results of (b) and (c) without using the axiom of choice. [For (b) use the Gram-Schmidt orthonormalization process; for (c) note that E is a ccc space.]

2. Suppose that E is a normed space over $\mathbf{R}$.

(a) If M is a subspace of E, $x \in E$, and $f : M \to \mathbf{R}$ is a continuous linear functional, show that f can be extended to a continuous linear functional $\bar{f}$ on the closed linear span of $M \cup \{x\}$ such that $\|\bar{f}\| = \|f\|$.

(b) If E has a dense well-orderable subcollection, show that f can be similarly extended to a continuous linear functional on E (*Hahn/Banach theorem*).

8 Lattices

What we now call the theory of lattices originated in the late nineteenth century from two sources: Dedekind came upon the notion of a *Dualgruppe*—what is now called a 'modular lattice'—in his investigation of the theory of divisibility; the other motivating example is the algebra of classes—theory of logical equivalence of propositions—which was studied by Boole, Schröder, and others. The work we shall be describing in this chapter is almost all the product of research in the 1930s by Birkhoff and Øre. Since then lattices have found natural application in functional analysis, projective geometry, set theory, logic, quantum mechanics, computer science, group theory, and number theory: 'Mathematical objects have lattices as dogs, fleas' (Isbell 1984, p. 94).

8.1 Lattices

Definition

A partially ordered collection $(L, \leq)$ is said to be a *lattice* if every finite subcollection of L has a supremum and an infimum.

It amounts to the same thing (remembering that $\varnothing$ is finite) to say that a partially ordered collection $(L, \leq)$ is a lattice if it satisfies:
(1) If $x, y \in L$ then $\{x, y\}$ has a supremum and an infimum in L.
(2) L has a greatest and a least element.
If $(L, \leq)$ is a lattice, then the supremum of $\{x, y\}$ is often denoted $x \sqcup y$ and the infimum is denoted $x \sqcap y$; the greatest element is denoted $\top$ and the least element is denoted $\bot$.

Remark 1 We shall call partially ordered collections satisfying condition (1) *pseudo-lattices*: a lattice is therefore a bounded psuedo-lattice. Many authors use 'lattice' for what we are calling pseudo-lattices.

Remark 2 The property of being a [pseudo-]lattice is self-dual.

Example 1 A totally ordered collection is a lattice iff it is bounded. In particular, if α is an ordinal then $(\boldsymbol{\alpha}, \leq)$ is a lattice unless α is a limit ordinal. $(\mathbf{1}, \leq)$, $(\mathbf{2}, \leq)$, and $(\mathbf{3}, \leq)$ are (up to isomorphism) the only lattices with fewer than four elements. $(\mathbf{1}, \leq)$ is called the *trivial* lattice: it is the only lattice in which $\bot = \top$.

Example 2 Every complete partially ordered collection is evidently a lattice.

Example 3 $(\omega, |)$ (example 3 of §4.2) is a lattice in which $m \sqcap n = \gcd(m, n)$ and $m \sqcup n = \text{lcm}(m, n)$.

Example 4 If A is a collection, then $\mathfrak{P}(A)$ is a lattice with respect to inclusion in which $X \sqcap Y = X \cap Y$ and $X \sqcup Y = X \cup Y$.

Abstract structures I have said that the theory of lattices originated in the late nineteenth century. But you should not expect to find in the writings of that time an abstract definition such as we have just given of a lattice as a set-theoretic structure satisfying certain axioms. It was not until the appearance of van der Waerden's *Moderne Algebra* in 1930 demonstrated to the mathematical community the power and elegance of abstract algebraic methods that the study of lattice theory as we now know it started in earnest. Before then it was explicit representations one dealt with in algebra.

Lemma 8.1.1

If $(L, \leq)$ is a lattice, then for all $x, y, z \in L$:

(L0) $x \sqcap x = x = x \sqcup x$ (Idempotency);

(L1) $x \sqcap y = y \sqcap x; x \sqcup y = y \sqcup x$ (Commutativity);

(L2) $(x \sqcap y) \sqcap z = x \sqcap (y \sqcap z); (x \sqcup y) \sqcup z = x \sqcup (y \sqcup z)$ (Associativity);

(L3) $x \sqcap (x \sqcup y) = x = x \sqcup (x \sqcap y)$ (Absorption);

(L4) $\bot \sqcap x = \bot; \bot \sqcup x = x$ (Zero);

(L5) $\top \sqcup x = \top; \top \sqcap x = x$ (Unit).

Proof Trivial. □

Lemma 8.1.2

If $(L, \leq)$ is a lattice, then for all $x, y \in L$:

$$x \leq y \Leftrightarrow x \sqcap y = x \Leftrightarrow x \sqcup y = y \quad \text{(Consistency)}.$$

Proof Trivial. □

Lemma 8.1.3

If $(L, \leq)$ is a lattice, then for all $x, y, z \in L$:

$$y \leq z \Rightarrow (x \sqcap y \leq x \sqcap z \text{ and } x \sqcup y \leq x \sqcup z) \quad \text{(Isotonicity)}.$$

Proof If $y \leq z$ then

$$\begin{aligned} x \sqcap y &= x \sqcap (y \sqcap z) \quad \text{[consistency]} \\ &= (x \sqcap x) \sqcap (y \sqcap z) \quad \text{(idempotency]} \\ &= (x \sqcap y) \sqcap (x \sqcap z) \quad \text{[comutativity and associativity]} \end{aligned}$$

and therefore

$$x \sqcap y \leq x \sqcap z \quad \text{[consistency]}.$$

The other inequality is proved similarly. □

Lemma 8.1.4

If $(L, \leq)$ is a lattice, then for all $x, y, z \in L$:
(a) $x \sqcup (y \sqcap z) \leq (x \sqcup y) \sqcap (x \sqcup z)$;
(b) $x \sqcap (y \sqcup z) \geq (x \sqcap y) \sqcup (x \sqcap z)$.

Proof (a) $y \sqcap z \leq y$, and therefore

$$x \sqcup (y \sqcap z) \leq x \sqcup y \quad \text{[isotonicity]}.$$

In the same way

$$x \sqcup (y \sqcap z) \leq x \sqcup z.$$

Consequently

$$x \sqcup (y \sqcap z) \leq (x \sqcup y) \sqcap (x \sqcup z).$$

(b) This is the dual of (a). □

Exercise Suppose that $(L, \sqcup, \sqcap)$ is a system consisting of a collection L endowed with two binary operations $(x, y) \mapsto x \sqcup y$ and $(x, y) \mapsto x \sqcap y$ satisfying conditions (L1) to (L3).
(a) Show that (L0) holds.
(b) Show that if we define a relation $\leq$ on L so that $x \leq y$ iff $x \sqcap y = x$ then $(L, \leq)$ is a pseudo-lattice in which the supremum [respectively infimum] of $\{x, y\}$ is $x \sqcup y$ [respectively $x \sqcap y$].
(c) If in addition L has elements $\perp$ and $\top$ satisfying (L4) and (L5), show that $(L, \leq)$ is a lattice in which the bottom [respectively top] is $\perp$ [respectively $\top$].

Throughout the remainder of this chapter L, L', L_1, L_2, etc. will be lattices.

8.2 Lattice homomorphisms

Lemma 8.2.1

If $f : L \to L'$ is a mapping then these three assertions are equivalent:
(i) f *is increasing;*
(ii) $f(x) \sqcup f(y) \leq f(x \sqcup y)$ *for all* $x, y \in L$;
(iii) $f(x) \sqcap f(y) \geq f(x \sqcap y)$ *for all* $x, y \in L$.

(i) ⇒ (ii) $x \leq x \sqcup y$ and so $f(x) \leq f(x \sqcup y)$. Similarly $y \leq x \sqcup y$ and so $f(y) \leq f(x \sqcup y)$. Consequently

$$f(x) \sqcup f(y) \leq f(x \sqcup y).$$

(ii) ⇒ (i) $x \leq y \Rightarrow y = x \sqcup y$

$$\Rightarrow f(y) = f(x \sqcup y) = f(x) \sqcup f(y)$$

$$\Rightarrow f(x) \leq f(y).$$

(i) ⇔ (iii) Similar. □

Definition

A mapping $f : L \to L'$ is called a *lattice homomorphism* if it satisfies:

$$f(\sup F) = \sup f[F]$$

$$f(\inf F) = \inf f[F]$$

for every finite subcollection F of L.

It evidently amounts to the same thing to say that $f : L \to L'$ is a lattice homomorphism iff it satisfies:

$$f(x \sqcap y) = f(x) \sqcap f(y) \quad \text{for all } x, y \in L;$$

$$f(x \sqcup y) = f(x) \sqcup f(y) \quad \text{for all } x, y \in L;$$

$$f(\bot) = \bot \quad \text{and} \quad f(\top) = \top.$$

Remark 1 If $f : L \to L'$ and $g : L' \to L''$ are lattice homomorphisms then $g \circ f : L \to L''$ is also a lattice homomorphism. Moreover, $\mathrm{id}_L : L \to L$ is trivially a lattice homomorphism.

Remark 2 It is clear from lemma 8.2.1 that every lattice homomorphism is increasing.

Remark 3 A mapping between lattices is an isomorphism iff it is a bijective lattice homomorphism.

Example 1 Figure 4 illustrates an increasing mapping f from one lattice to another which is not a lattice homomorphism since

$$f(a) \sqcup f(b) = \bot < \top = f(a \sqcup b).$$

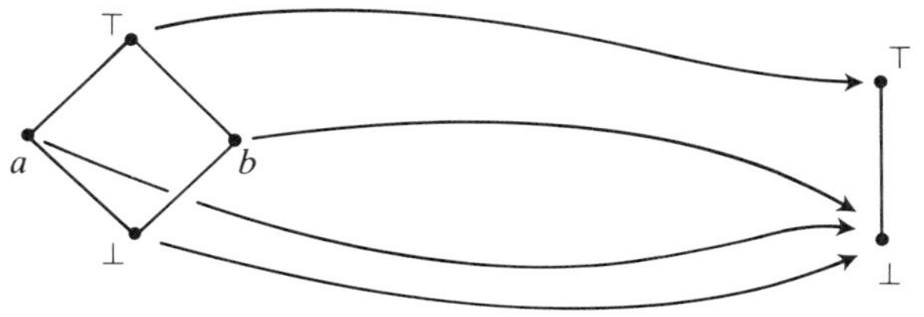

Figure 4

Lemma 8.2.2

If $M \subseteq L$, then $\mathrm{id}_M : M \to L$ is a lattice homomorphism from $(M, \leq_M)$ to $(L, \leq)$ iff M satisfies the following:

$$x, y \in M \Rightarrow x \sqcup y \in M;$$

$$x, y \in M \Rightarrow x \sqcap y \in M;$$

$$\bot, \top \in M.$$

Proof Straightforward. □

Definition

$(M, \leq_M)$ is called a *sublattice of* $(L, \leq)$ if it satisfies the equivalent conditions of lemma 8.2.2.

Remark 4 It is easy to check that the intersection of a family of sublattices of L is itself a sublattice of L, although the union even of two sublattices need not be one. So the sublattices of L form a complete lattice Sub(L) with respect to inclusion: the greatest element of Sub(L) is L itself and the least element is $\{\top, \bot\}$.

We say that $f : L \to L'$ is a *lattice embedding* if it is an embedding and $f[L]$ is a sublattice of L'; this is the case iff $f : L \to L'$ is an injective lattice homomorphism.

Definition

If $A \subseteq L$ then the intersection of all the sublattices of L containing A is denoted $[A]$.

It follows that $[A]$ is the smallest sublattice of L containing A: for this reason it is said to be *generated by A*.

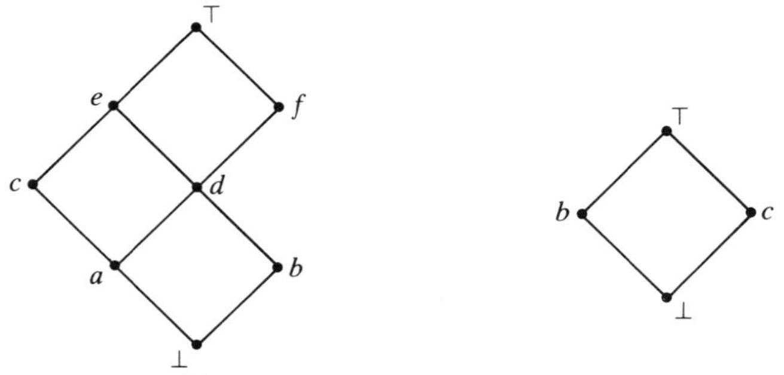

(a) (b)

Figure 5

Remark 5 It is possible for a subcollection M of L to be a lattice with respect to the induced partial ordering $\leq_M$ without being a sublattice of L. For an example consider a lattice L which looks like Fig. 5(a). The subcollection $M = \{\bot, b, c, \top\}$ looks like Fig. 5(b), but $b \sqcup_M c = \top$ whereas $b \sqcup_L c = e$.

Lemma 8.2.3

If s is an equivalence relation on L, then the canonical quotient mapping $p_s : L \to L/r$ is a lattice homomorphism from $(L, \leq)$ onto $(L/s, \leq/s)$ iff

$$(x \; s \; y \; and \; z \; s \; t) \Rightarrow (x \sqcap z \; s \; y \sqcap t \; and \; x \sqcup z \; s \; y \sqcup t)$$

for all $x, y, z, t \in L$.

Proof Straightforward. □

Definition

In the circumstances described in lemma 8.2.3 s is said to be a *congruence* on L and the pair $(L/s, \leq/s)$ is called a *quotient lattice* of $(L, \leq)$.

Example 2 If $f : L \to L'$ is a lattice homomorphism, then $\mathrm{Ker}[f]$ is a congruence on L.

Remark 6 It is easy to check that the intersection of a family of congruences on L is itself a congruence on L. So the congruences form a complete lattice $\mathrm{Cong}(L)$ with respect to inclusion: the greatest element of $\mathrm{Cong}(L)$ is the *improper* congruence $L \times L$; the least element is the *trivial* congruence id_L. The quotient lattice is trivial iff the congruence is improper.

Definition

If $A, B \subseteq L$ then the intersection of all the congruences on L such that A is contained in the equivalence class of $\bot$ and B is contained in the equivalence class of $\top$ is denoted $[A, B]$.

So $[A, B]$ is the smallest congruence on L which relates all the elements of A to $\top$ and all the elements of B to $\bot$.

Proposition 8.2.4

If $(L_i)_{i \in I}$ is a family of lattices then $\Pi_{i \in I} L_i$ is a lattice with the product partial ordering and all the projections $\mathrm{pr}_j : \Pi_{x \in I} L_i \to L_j$ are lattice homomorphisms.

Proof Straightforward. □

Exercise Define *pseudo-lattice homomorphism*, *sub-pseudo-lattice*, and *quotient pseudo-lattice*; check which of the statements in this section generalize to these notions.

8.3 Distributive lattices

We shall restrict ourselves from now on to considering those lattices in which the inequalities of lemma 8.1.4 are in fact equalities; such lattices will be called 'distributive'.

Definition

L is said to be *distributive* if it satisfies:

$$x \sqcap (y \sqcup z) = (x \sqcap y) \sqcup (x \sqcap z) \quad \text{for all } x, y, z \in L;$$

$$x \sqcup (y \sqcap z) = (x \sqcup y) \sqcap (x \sqcup z) \quad \text{for all } x, y, z \in L.$$

Theorem 8.3.1 (Bergmann 1929)

There six assertions are equivalent:
(*i*) L *is distributive;*
(*ii*) $x \sqcap (y \sqcup z) \leq (x \sqcap y) \sqcup (x \sqcap z)$ *for all* $x, y, z \in L$;
(*iii*) $x \sqcup (y \sqcap z) \geq (x \sqcup y) \sqcap (x \sqcup z)$ *for all* $x, y, z \in L$;
(*iv*) $(x \sqcup y) \sqcap z \leq x \sqcup (y \sqcap z)$ *for all* $x, y, z \in L$;
(*v*) $(x \sqcap z = y \sqcap z$ *and* $x \sqcup z = y \sqcup z) \Rightarrow x = y$ *for all* $x, y, z \in L$;
(*vi*) $(x \sqcup y) \sqcap (y \sqcup z) \sqcap (z \sqcup x) = (x \sqcap y) \sqcup (y \sqcap z) \sqcup (z \sqcap x)$ *for all* $x, y, z \in L$.

(i)$\Leftrightarrow$((ii) and (iii)) Immediate [lemma 8.1.4].
(iii)$\Rightarrow$(iv) Suppose (iii) holds. Then

$$\begin{aligned}(x \sqcup y) \sqcap z &\leq (x \sqcup y) \sqcap (x \sqcup z) \quad \text{[isotonicity]}\\ &\leq x \sqcup (y \sqcap z) \quad \text{[(iii)]}.\end{aligned}$$

(ii) $\Rightarrow$ (iv) This is the dual of (iii) $\Rightarrow$ (iv) since (iv) is self-dual [commutativity].
(iv)$\Rightarrow$(iii) Suppose (iv) holds. Then

$$\begin{aligned}(x \sqcup y) \sqcap (x \sqcup z) &\leq x \sqcup (y \sqcap (x \sqcup z)) \quad \text{[iv)]}\\ &= x \sqcup ((x \sqcup z) \sqcap y) \quad \text{[commutativity]}\\ &\leq x \sqcup (x \sqcup (z \sqcap y)) \quad \text{[(iv) and isotonicity]}\\ &= x \sqcup (x \sqcup (y \sqcap z)) \quad \text{[commutativity]}\\ &= (x \sqcup x) \sqcup (y \sqcap z) \quad \text{[associativity]}\\ &= x \sqcup (y \sqcap z) \quad \text{[idempotency]}.\end{aligned}$$

(iv)$\Rightarrow$(ii) This is the dual of (iv)$\Rightarrow$(iii).
(i)$\Rightarrow$(v) Suppose that L is distributive. If $x, y, z \in L$ are such that $x \sqcap z = y \sqcap z$ and $x \sqcup z = y \sqcup z$, then

$$\begin{aligned}x &= x \sqcap (x \sqcup z) \quad \text{[absorption]}\\ &= x \sqcap (y \sqcup z)\end{aligned}$$

$$= (x \sqcap y) \sqcup (x \sqcap z) \quad [\text{(i)}]$$
$$= (x \sqcap y) \sqcup (y \sqcap z)$$
$$= (y \sqcap x) \sqcup (y \sqcap z) \quad [\text{commutativity}]$$
$$= y \sqcap (x \sqcup z) \quad [\text{(i)}]$$
$$= y \sqcap (y \sqcup z)$$
$$= y \quad [\text{absorption}].$$

(v) $\Rightarrow$ (iii) Suppose that (v) holds. We intend first to show that for all $a, b, c \in L$

$$(a \sqcap b) \sqcup (a \sqcap c) = a \sqcap (b \sqcup (a \sqcap c)). \tag{8.3.1}$$

If we let

$$u = (a \sqcap b) \sqcup (a \sqcap c)$$
$$v = a \sqcap (b \sqcup (a \sqcap c))$$

then certainly $u \leq v$ [lemma 8.1.4(b)] and so

$$b \sqcup u \leq b \sqcup v \quad [\text{isotonicity}].$$

But also $b \leq b \sqcup (a \sqcap c)$ and so

$$b \sqcup v \leq (b \sqcup (a \sqcap c)) \sqcup v \quad [\text{isotonicity}]$$
$$= (b \sqcup (a \sqcap c)) \sqcup (a \sqcap (b \sqcup (a \sqcap c)))$$
$$= b \sqcup (a \sqcap c) \quad [\text{absorption}]$$
$$= (b \sqcup (a \sqcap b)) \sqcup (a \sqcap c) \quad [\text{absorption}]$$
$$= b \sqcup u \quad [\text{associativity}].$$

Hence $b \sqcup u = b \sqcup v$. A dual argument shows that $b \sqcap u = b \sqcap v$. Therefore by (v) $u = v$, which establishes (8.3.1). Next notice that

$$c \leq a \Rightarrow (a \sqcap b) \sqcup c = (a \sqcap b) \sqcup (a \sqcap c) \quad [\text{consistency}]$$
$$= a \sqcap (b \sqcup (a \sqcap c)) \quad [(8.3.1)]$$
$$= a \sqcap (b \sqcup c) \quad [\text{consistency}]. \tag{8.3.2}$$

Suppose now that $x, y, z \in L$ and let

$$a = (x \sqcap (y \sqcup z)) \sqcup (y \sqcap z),$$
$$b = (y \sqcap (z \sqcup x)) \sqcup (z \sqcap x),$$
$$c = (z \sqcap (x \sqcup y)) \sqcup (x \sqcap y).$$

Then

$$a \sqcup b = (x \sqcap (y \sqcup z)) \sqcup (y \sqcap z) \sqcup (y \sqcap (z \sqcup x)) \sqcup (z \sqcap x)$$
$$= (x \sqcap (y \sqcup z)) \sqcup (y \sqcap (z \sqcup x))$$

$$= (x \sqcup (y \sqcap (x \sqcup z))) \sqcap (y \sqcup z) \quad [(8.3.2)]$$
$$= (x \sqcup y) \sqcap (x \sqcup z) \sqcap (y \sqcup z) \quad [(8.3.1)]. \qquad (8.3.3)$$

By cycling x, y, and z in this argument we can see that

$$a \sqcup b = b \sqcup c = c \sqcup a$$

and dually

$$a \sqcap b = b \sqcap c = c \sqcap a.$$

Hence by condition (v) $a = b = c$. Therefore

$$(x \sqcup (y \sqcap z)) \sqcap (y \sqcup z) = (x \sqcup (y \sqcap z)) \sqcap (y \sqcup z) \quad [(8.3.2)]$$
$$= a = a \sqcup b$$
$$= ((x \sqcup y) \sqcap (x \sqcup z)) \sqcap (y \sqcup z) \quad [(8.3.3)]. \qquad (8.3.4)$$

Also

$$(x \sqcup (y \sqcap z)) \sqcup (y \sqcup z) = x \sqcup y \sqcup z \quad [\text{absorption}]$$
$$= (x \sqcup (y \sqcap (x \sqcup z))) \sqcup y \sqcup z \quad [\text{absorption}]$$
$$= ((x \sqcup y) \sqcap (x \sqcup z)) \sqcup (y \sqcup z) \quad [(8.3.1)]. \qquad (8.3.5)$$

It follows by condition (v) from (8.3.4) and (8.3.5) that

$$x \sqcup (y \sqcap z) = (x \sqcup y) \sqcap (x \sqcup z).$$

(i) $\Rightarrow$ (vi) Suppose that L is distributive. Then

$$(x \sqcup y) \sqcap (x \sqcup z) \sqcap (y \sqcup z) = (x \sqcup (y \sqcap z)) \sqcap (y \sqcup z)$$
$$= (x \sqcap (y \sqcup z)) \sqcup ((y \sqcap z) \sqcap (y \sqcup z))$$
$$= (x \sqcap (y \sqcup z)) \sqcup (y \sqcap z)$$
$$= (x \sqcap y) \sqcup (x \sqcap z) \sqcup (y \sqcap z).$$

(vi) $\Rightarrow$ (v) Suppose that (vi) holds. If $x, y, z \in L$ are such that $x \sqcap z = y \sqcap$ and $x \sqcup z = y \sqcup z$, then

$$x = x \sqcap (x \sqcup z) \sqcap (x \sqcup y)$$
$$= x \sqcap (x \sqcup z) \sqcap (y \sqcup z) \sqcap (x \sqcup y)$$
$$= x \sqcap ((x \sqcap z) \sqcup (y \sqcap z) \sqcup (x \sqcap y))$$
$$= x \sqcap ((x \sqcap z) \sqcup (x \sqcap y))$$
$$= (x \sqcap z) \sqcup (x \sqcap y).$$

Similarly

$$y = (y \sqcap z) \sqcup (x \sqcap y),$$

whence $x = y$. □

xample 1 The lattice $(\omega, |)$ is distributive. For if $\mathrm{lcm}(m,r) = \mathrm{lcm}(n,r)$ and $\gcd(m,r) = \gcd(n,r)$, then either $r = 0$, in which case

$$m = \gcd(m,0) = \gcd(n,0) = n,$$

or $r \neq 0$, in which case

$$\begin{aligned} mr &= \mathrm{lcm}(m,r)\gcd(m,r) \quad [(4.2.7)] \\ &= \mathrm{lcm}(n,r)\gcd(n,r) \\ &= nr \quad [(4.2.7)] \end{aligned}$$

and therefore $m = n$.

xample 2 The five-element lattices represented by Fig. 6(a)–(c) are all distributive.

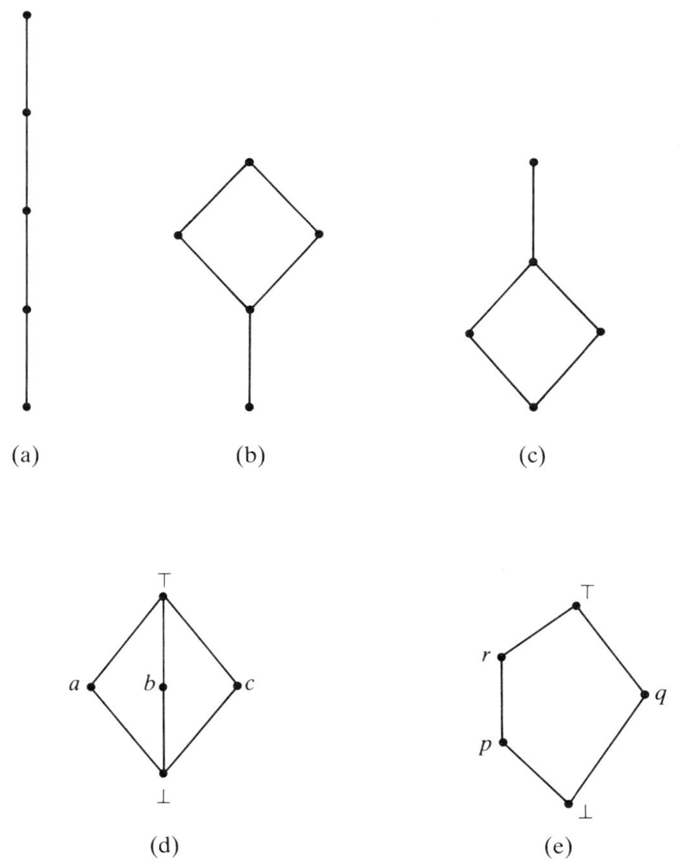

Figure 6

Example 3 The set of normal subgroups of a group G is, when partially ordered by inclusic a lattice in which $H \sqcap K = H \cap K$ and $H \sqcup K = HK$. This lattice is not in gener distributive: the simplest counterexample is the non-cyclic group of order 4 (t Klein *Viergruppe*), whose lattice of normal subgroups looks like Fig. 6(d). Her

$$a \sqcap (b \sqcup c) = a > \bot = (a \sqcap b) \sqcup (a \sqcap c),$$

and so the lattice is not distributive.

Example 4 Consider a lattice which looks like Fig. 6(e). Here

$$(p \sqcup q) \sqcap r = r > p = p \sqcup (q \sqcap r),$$

and so the lattice is not distributive.

Proposition 8.3.2

If L is distributive and $A \subseteq L$ then

$$[A] = \left\{x \in L : x = \sup_{B \in \mathscr{B}} \inf B \text{ for some } \mathscr{B} \in \mathfrak{F}(\mathfrak{F}(A))\right\}.$$

Proof Exercise. [

Exercises

1. (a) Show that every sublattice and every quotient lattice of a distributive latti is distributive.
   (b) Show that a product lattice is distributive if all the factor lattices a distributive.
2. Prove proposition 8.3.2.
3. Prove (8.3.7).
4. Show that a lattice L has property (v) of theorem 8.3.1 iff it does not have sub-psuedo-lattice isomorphic to either of the lattices in examples 3 and above.

8.4 Heyting, Brouwerian, and Boolean lattices

Definition

If $a, b \in L$, then we let

$$a \mathrel{-\!\!\ni} b = \max\{x \in L : x \sqcap a \leq b\}$$

$$a \mathrel{\in\!\!-} b = \min\{x \in L : x \sqcup a \geq b\}$$

whenever these objects exist.

So $a \mathrel{-\!\!\ni} b$ and $a \mathrel{\in\!\!-} b$ are defined (if they exist) by the equivalences

$$x \leq a \mathrel{-\!\!\ni} b \Leftrightarrow x \sqcap a \leq b$$

$$x \geq a \mathrel{\in\!\!-} b \Leftrightarrow x \sqcup a \geq b.$$

efinition

If $a \in L$ we let

$$a^* = a \multimap \bot$$

$$a_* = a \multimapinv \top$$

whenever these exist.

So a^* and a_* are defined (if they exist) by the equivalences

$$x \leq a^* \Leftrightarrow x \sqcap a = \bot$$

$$x \geq a_* \Leftrightarrow x \sqcup a = \top.$$

efinition

L is said to be *Heyting* [respectively *Brouwerian*] if $a \multimap b$ [respectively $a \multimapinv b$] exists for all $a, b \in L$.

emark 1 The properties of being Heyting and Brouwerian are evidently duals of one another.

roposition 8.4.1

Every Heyting [respectively Brouwerian] lattice is distributive.

roof Suppose that L is Heyting and $x, y, z \in L$. Let $a = (x \sqcap y) \sqcup (x \sqcap z)$. Then plainly $y \sqcap x \leq a$ and so $y \leq x \multimap a$. But similarly $z \sqcap x \leq a$ and so $z \leq x \multimap a$. Hence $y \sqcup z \leq x \multimap a$, and therefore $x \sqcap (y \sqcup z) \leq a$. It follows that L is distributive [theorem 8.3.1]. The case when L is Brouwerian now follows by duality. □

xercise 1. Show that every finite distributive lattice is Heyting and Brouwerian.

efinition

Elements $a, a' \in L$ are said to be *complementary* if $a \sqcap a' = \bot$ and $a \sqcup a' = \top$. The collection of elements of L which have a complement is written $\mathfrak{B}(L)$.

emma 8.4.2

Let L be distributive. If $a, a' \in L$ are complementary then

(a) $a \multimap b = b \sqcup a'$ *and* $a \multimapinv b = b \sqcap a'$ *for all* $b \in L$;
(b) $a^* = a' = a_*$.

roof If $x \leq b \sqcup a'$ then

$$\begin{aligned} a \sqcap x &\leq a \sqcap (b \sqcup a') \\ &= (a \sqcap b) \sqcup (a \sqcap a') \end{aligned}$$

$$= (a \sqcap b) \sqcup \bot$$

$$= a \sqcap b$$

$$\leq b.$$

And if $a \sqcap x \leq b$ then

$$\begin{aligned} x &= x \sqcap \top \\ &= x \sqcap (a \sqcup a') \\ &= (x \sqcap a) \sqcup (x \sqcap a') \\ &\leq b \sqcup a'. \end{aligned}$$

So $a \dashv b = b \sqcup a'$. In particular,

$$a^* = a \dashv \bot = \bot \sqcup a' = a'.$$

The other properties follow by duality. □

Remark 2 If L is distributive, then an element a of L has at most one complemen which we shall from now on invariably denote by a^*. It is not tru conversely that if every element of L has a unique complement then L i distributive (Dilworth 1945), but no elementary counterexample seems t be known.

Definition

L is said to be *Boolean* if it is distributive and every element of L has complement.

Proposition 8.4.3

Every Boolean lattice is Heyting and Brouwerian.

Proof Immediate [lemma 8.4.2(a)]. □

Example 1 $\mathfrak{P}(A)$ is a Boolean lattice with respect to inclusion: the complement of a subcollec tion B of A is $A - B$. The particular case in which $A = \mathbf{D}$ (so that $\mathfrak{P}(A)$ is th collection of all classes) has been especially important historically.

Example 2 The sublattice $\mathfrak{F}(A)$ of $\mathfrak{P}(A)$ is not Boolean unless A is finite: the Boolean sublattic of $\mathfrak{P}(A)$ which it generates is the collection $\mathfrak{FC}(A)$ of finite and cofinite subcollec tions of A. (B is *cofinite* in A if $A - B$ is finite.) $\mathfrak{FC}(A)$ is countable iff A is countable

Example 3 Any bounded totally ordered collection $(A, \leq)$ is both a Heyting and a Brou werian lattice. Its operations are as follows:

$$a \sqcup b = \max(a, b)$$

$$a \sqcap b = \min(a, b)$$

$$a \dashv b = \begin{cases} \top & \text{if } a \leq b \\ b & \text{otherwise} \end{cases}$$

$$a \in b = \begin{cases} \bot & \text{if } a \geq b \\ b & \text{otherwise} \end{cases}$$

$$a^* = \begin{cases} \top & \text{if } a = \bot \\ \bot & \text{otherwise} \end{cases}$$

$$a_* = \begin{cases} \bot & \text{if } a = \top \\ \top & \text{otherwise.} \end{cases}$$

But $(A, \leq)$ is not Boolean if it has more than two elements.

xample 4 A lattice which is Brouwerian but not Heyting cannot be finite (see exercise 1). One of the simplest examples is $(\omega, |)$, in which

$$m \in n = \begin{cases} \dfrac{n}{\gcd(m, n)} & \text{if } m \neq 0 \quad \text{or} \quad n \neq 0 \\ 1 & \text{if } m = n = 0 \end{cases}$$

$$m_* = \begin{cases} 1 & \text{if } m = 0 \\ 0 & \text{otherwise.} \end{cases}$$

Suppose for a moment that 2^* exists. Then

$$n | 2^* \Leftrightarrow \gcd(n, 2) = 1 \Leftrightarrow n \text{ is odd;}$$

but the only number divisible by all odd numbers is 0, which is even. Contradiction. So in particular $(\omega, |)$ is not Heyting. To get an example of a Heyting lattice which is not Brouwerian just turn this one upside down.

xample 5 If V is a (finite-dimensional) vector space, then the subspaces of V form a complete lattice with respect to inclusion: the infimum of a collection $\{U_i : i \in I\}$ of subspaces of V is their intersection $\bigcap_{i \in I} U_i$; the supremum is their linear span $\Sigma_{i \in I} U_i$. Moreover, every element of this lattice has a complement. However, complements are not in general unique: in $\mathbf{R}^2$, for instance, the x-axis is complementary to every other line through the origin. This indicates, of course, that the lattice is not distributive.

xercises

2. If $f : L \to L'$ is a lattice homomorphism and x is a complemented element of L, show that $f(x)$ is a complemented element of L' and $f(x^*) = f(x)^*$.
3. (a) Show that a sublattice M of a Boolean lattice L is Boolean iff $x \in M \Rightarrow x^* \in M$.
   (b) Show that every quotient lattice of a distributive lattice is distributive.
   (c) Show that a product lattice is Boolean iff all the factor lattices are Boolean.
   (d) Show that if L is distributive then $\mathfrak{B}(L)$ is its largest Boolean sublattice.

It is often helpful to bear in mind that Boolean lattices are self-dual: to be less precise, if you turn one upsidedown, it looks the same. To be more precise:

Proposition 8.4.4

If L is Boolean then the function $x \mapsto x^*$ is the graph of an isomorphism of L onto its dual.

Proof This follows at once from the fact that in a Boolean lattice $x^{**} = x$ and $x \leq y \Leftrightarrow y^* \leq x^*$. □

Proposition 8.4.5

If L is Boolean then the Boolean sublattice of L generated by $A \subseteq L$ is $[A \cup A^*]$ (where $A^* = \{x^* : x \in A\}$).

Proof Easy. □

Exercises

4. Show that there are (up to isomorphism) no five-element lattices apart from the five illustrated in Fig. 6.
5. Suppose that L is Heyting.
   (a) Let $\mathfrak{R}(L) = \{x \in L : x^{**} = x\}$. Show that $\mathfrak{B}(L) \subseteq \mathfrak{R}(L) \subseteq L$ and that $\mathfrak{R}(L)$ is a Boolean lattice with respect to the induced partial order relation (but not necessarily a sublattice of L: see (d)). [*Hint* The infimum in $\mathfrak{R}(L)$ of $\{x, y\}$ is $x \sqcap y$ but the supremum is $(x \sqcup y)^{**}$.]
   (b) Show that these four assertions are equivalent:
      (i) L is Boolean;
      (ii) $\mathfrak{R}(L) = L$;
      (iii) $x \sqcup x^* = \top$ for all $x \in L$;
      (iv) $\mathfrak{B}(L) = L$.
   (c) Show these four assertions are equivalent:
      (i) $\mathfrak{R}(L) = \mathfrak{B}(L)$;
      (ii) $x^* \sqcup x^{**} = \top$ for all $x \in L$;
      (iii) $(x \sqcap y)^* = x^* \sqcup y^*$ for all $x, y \in L$;
      (iv) $x \sqcap y = \bot \Rightarrow x^* \sqcup y^* = \top$ for all $x, y \in L$;
      (v) $\mathfrak{R}(L)$ is a Boolean sublattice of L.
      L is said to be *Stonian* if it satisfies these conditions.
   (d) Show that the lattice illustrated in Fig. 5(a) is Heyting but not Stonian.
6. A *frame* is a complete Heyting lattice.
   (a) Show that L is a frame iff it is a complete lattice in which $a \sqcap \sup B = \sup_{b \in B}(a \sqcap b)$ for every $a \in L$ and $B \subseteq L$.
   (b) Show that if L is a frame then $\mathfrak{R}(L)$ is a complete Boolean lattice.
7. Show that every partially well ordered Boolean lattice is finite.

8.5 Free Boolean lattices

In this section we shall show how to construct from an arbitrary family $(\mathbf{X}_i)_{i \in I}$ of objects a Boolean lattice Bool(I) which contains and is generated by $\{\mathbf{X}_i : i \in I\}$ and, moreover, is such that the only Boolean equations which hold between the $\mathbf{X}_i$ are those which hold in all Boolean lattices. Such a Boolean lattice is said to be 'free'.

Lemma 8.5.1

If L is a Boolean lattice generated by a subcollection A then

$\operatorname{card}(L) \leq 2^{4^{\operatorname{card}(A)}}$.

Proof This follows easily from proposition 8.4.5. □

Lemma 8.5.2

If L is a Boolean lattice generated by a subcollection A then any two lattice homomorphisms $L \to M$ which agree on A agree on L.

Proof Exercise. □

Theorem 8.5.3

For each collection I there exists a definite Boolean lattice Bool(I) *containing a family $(\mathbf{X}_i)_{i \in I}$ such that if B is any Boolean lattice and $(a_i)_{i \in I}$ is a family in B indexed by I then there is exactly one lattice homomorphism $u : \text{Bool}(I) \to B$ such that $u(\mathbf{X}_i) = a_i$ for all $i \in I$.*

Proof Let $((B_\kappa, \leq_\kappa, (x_{\kappa,i})_{i \in I}))_{\kappa \in K}$ be a family consisting of all ordered triples $(B_\kappa, \leq_\kappa, (x_{\kappa,i})_{i \in I})$ such that $(B_\kappa, \leq_\kappa)$ is a Boolean lattice, $B_\kappa \subseteq \mathfrak{P}(\mathfrak{P}(I \cup I))$, and $(x_{\kappa,i})_{i \in I}$ is a family in B. Now the product lattice $\Pi_{\kappa \in K} B_\kappa$ is Boolean (cf. exercise 2(c) of §8.4). Let $\mathbf{X}_i = (x_{\kappa,i})_{\kappa \in K}$ for each $i \in I$ and let Bool(I) be the Boolean sublattice of $\Pi_{\kappa \in K} B_\kappa$ generated by $\{\mathbf{X}_i : i \in I\}$.

Suppose now that B is a Boolean lattice and $(a_i)_{i \in I}$ is a family in B. If C is the Boolean sublattice of B generated by $\{a_i : i \in I\}$ then

$$\begin{aligned}\operatorname{card}(C) &\leq 2^{4^{\operatorname{card}(I)}} \quad \text{[lemma 8.5.1]}\\ &= \operatorname{card}(\mathfrak{P}(\mathfrak{P}(I \cup I))).\end{aligned}$$

So for some $\kappa_0 \in K$ there is a lattice isomorphism $g : B_{\kappa_0} \to C$ such that $g(x_{\kappa_0,i}) = a_i$ for all $i \in I$. If we let $u = g \circ \text{pr}_{\kappa_0} | \text{Bool}(I)$ then u is the graph of a lattice homomorphism Bool$(I) \to B$ such that

$$u(\mathbf{X}_i) = g(\text{pr}_{\kappa_0}(X_i)) = g(x_{\kappa_0,i}) = a_i \quad \text{for all } i \in I.$$

Moreover, the uniqueness of u follows at once from lemma 8.5.2 since Bool(I) is generated by $\{\mathbf{X}_i : i \in I\}$. □

Definition

The Boolean lattice Bool(I) described in theorem 8.5.3 is called the *free Boolean lattice* on I and its elements are called *Boolean polynomials* in the *indeterminates* $\mathbf{X}_i$ $(i \in I)$. Bool(I) is sometimes denoted Bool$[\mathbf{X}_i]_{i \in I}$ to emphasize the role of the indeterminates.

Proposition 8.5.4

If $\mathrm{Bool}'[\mathbf{X}'_i]_{i \in I}$ *is another Boolean lattice with the property describe in theorem* 8.5.3 *then there exists exactly one lattice isomorphis* $u : \mathrm{Bool}[\mathbf{X}_i]_{i \in I} \to \mathrm{Bool}'[\mathbf{X}'_i]_{i \in I}$ *such that* $u(\mathbf{X}_i) = \mathbf{X}'_i$ *for all* $i \in I$.

Proof Exercise. [

Remark 1 Each Boolean polynomial f in the indeterminates $(\mathbf{X}_i)_{i \in I}$ gives rise in a obvious way for each Boolean lattice B to a mapping $f_B : {}^I B \to B$ define by letting $f_B((x_i)_{i \in I}) = u_B(f)$ where $u_B : \mathrm{Bool}(I) \to B$ is the unique lattic homomorphism such that $u_B(\mathbf{X}_i) = x_i$ for all $i \in I$.

Remark 2 If f and g are Boolean polynomials in the indeterminates $\mathbf{X}_i$ $(i \in I)$ an $f = g$ then for any Boolean lattice B and any family $(x_i)_{i \in I}$ in B w have $f_B((x_i)_{i \in I}) = g_B((x_i)_{i \in I})$. This is a precise statement of the claim w made at the beginning of this section that Boolean equations which hol between the inteterminates of a free Boolean lattice hold in general.

Remark 3 Constructions analogous to this one can be carried out in all branches o algebra. For example, if F is a field then ${}^I F$ has an obvious (pointwise vector space structure over F. It is easy to see that ${}^{(I)}F$ is a subspace an that if for each $i \in I$ we let $\mathrm{e}_i = (\delta_{ij})_{j \in I}$ then $\{\mathrm{e}_i : i \in I\}$ is a basis for ${}^{(I)}F$ Moreover, if V is another vector space over F and $(a_i)_{i \in I}$ is a family in (not necessary a basis) then there is a unique linear mapping $f : {}^{(I)}F \to$ such that $f(\mathrm{e}_i) = a_i$ for all $i \in I$. ${}^{(I)}F$ is therefore the *free vector space* on

Example $\mathrm{Bool}[\mathbf{X}, \mathbf{Y}]$ looks like Fig. 7.

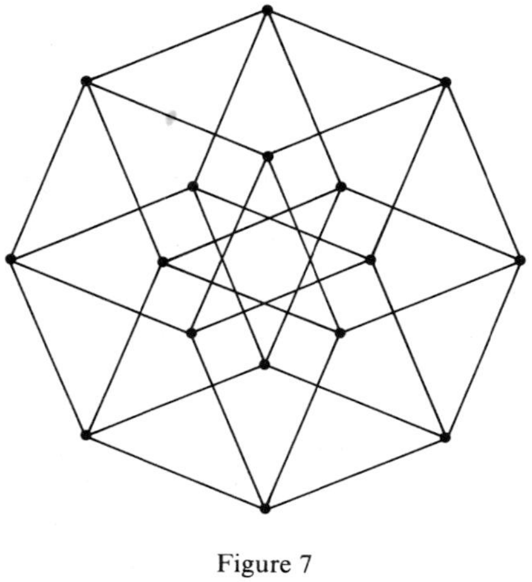

Figure 7

xercises

1. (a) Prove lemma 8.5.2.
   (b) Prove proposition 8.5.4.
2. Show that if I is finite then $\mathrm{card}(\mathrm{Bool}(I)) = 2^{2^{\mathrm{card}(I)}}$ and that if I is infinite and well-orderable then $\mathrm{card}(\mathrm{Bool}(I)) = \mathrm{card}(I)$.
3. Show that $\mathrm{Bool}(\boldsymbol{\omega})$ is (up to isomorphism) the only countable Boolean lattice without atoms.

.6 Spatial lattices

In this section L is a distributive lattice and $A \subseteq L$.

efinition

$\mathrm{U}_A^{\downarrow}(x) = A \cap \downarrow x$ and $\mathrm{U}_A^{\uparrow}(x) = A - \uparrow x$.

We shall devote the remainder of this chapter to a study of the mappings $\mathrm{U}_A^{\downarrow} : L \to \mathfrak{P}(A)$ and $\mathrm{U}_A^{\uparrow} : L \to \mathfrak{P}(A)$ thus defined. In particular, we should like to know in what circumstances they are lattice embeddings. We shall concentrate for the sake of convenience on $\mathrm{U}_A^{\downarrow}$: the properties of $\mathrm{U}_A^{\uparrow}$ are similar.

We start by noting the following properties, all of which are easy to verify:

$$\mathrm{U}_A^{\downarrow}(x) \cap \mathrm{U}_A^{\downarrow}(y) = \mathrm{U}_A^{\downarrow}(x \sqcap y); \tag{8.6.1}$$

$$\mathrm{U}_A^{\downarrow}(x) \cup \mathrm{U}_A^{\downarrow}(y) \subseteq \mathrm{U}_A^{\downarrow}(x \sqcup y); \tag{8.6.2}$$

$$\mathrm{U}_A^{\downarrow}(\top) = A. \tag{8.6.3}$$

Now we want to obtain a condition on A which will ensure that $U_A^{\downarrow}$ is a lattice homomorphism.

efinition

(1) An element $a \in L$ is *irreducible* if

$$a = \sup F \Rightarrow a \in F \quad \text{for every finite } F \subseteq L;$$

the collection of all irreducible elements of L is denoted $\mathrm{Irr}(L)$.

(2) Dually, an element $a \in L$ is *coirreducible* if

$$a = \inf F \Rightarrow a \in F \quad \text{for every finite } F \subseteq L;$$

the collection of all coirreducible elements of L is denoted $\mathrm{Pt}(L)$.

It amounts to the same thing to say that $a \in L$ is irreducible if it satisfies:

$$a = x \sqcup y \Rightarrow (a = x \text{ or } a = y); \tag{8.6.4}$$

$$a \neq \bot. \tag{8.6.5}$$

Moreover it is a straightforward exercise to show that (8.6.4) is equivalent to the apparently weaker condition

$$a \leq x \sqcup y \Rightarrow (a \leq x \text{ or } a \leq y). \quad (8.6.4$$

We leave it to the reader to formulate the dual description of the coirreducible elements.

Remark It is easy to see that every [co]atom of L is [co]irreducible. However, the converse does not hold in general.

Example 1 If $(B, \leq)$ is a bounded totally ordered collection then every element of B other than $\perp$ is irreducible, but there cannot be more than one atom and there need not be any.

Proposition 8.6.1

If L is Boolean then every [co]irreducible element of L is a [co]atom.

Proof Suppose that $a \in L$ is irreducible. If $b \leq a$, then

$$b \sqcup (b^* \sqcap a) = (b \sqcup b^*) \sqcap (b \sqcup a) = \top \sqcap (b \sqcup a) = b \sqcap a = a;$$

so by hypothesis either $b = a$ or $b^* \sqcap a = a$, and in the latter case $b \leq a \leq b^*$, so that $b = b \sqcap b^* = \perp$. This shows that a is an atom. The dual assertion follows by duality. □

The significance of irreducibles in our current discussion is that they provide the criterion for deciding whether $\mathrm{U}_A^{\downarrow} : L \to \mathfrak{P}(A)$ is a lattice homomorphism.

Proposition 8.6.2

$\mathrm{U}_A^{\downarrow} : L \to \mathfrak{P}(A)$ [respectively $\mathrm{U}_A^{\uparrow} : L \to \mathfrak{P}(A)$] is a lattice homomorphism iff $A \subseteq \mathrm{Irr}(L)$ [respectively $A \subseteq \mathrm{Pt}(L)$].

Proof It is clear, bearing in mind (8.6.1)–(8.6.3), that $\mathrm{U}_A^{\downarrow} : L \to \mathfrak{P}(A)$ is a lattice homomorphism iff

$$\mathrm{U}_A^{\downarrow}(x \sqcup y) \subseteq \mathrm{U}_A^{\downarrow}(x) \cup \mathrm{U}_A^{\downarrow}(y)$$
$$\mathrm{U}_A^{\downarrow}(\perp) = \varnothing,$$

i.e. iff (8.6.4′) and (8.6.5) are satisfied by all the elements of A. □

Definition

(1) The collection A is a *base* for L if whenever $x \not\leq y$ in L there exists $a \in \mathrm{A}$ such that $a \leq x$ and $a \not\leq y$.
(2) A is a *cobase* for L if it is a base for L^{op}.

roposition 8.6.3

$U_A^{\downarrow}: L \to \mathfrak{P}(A)$ [*respectively* $U_A^{\uparrow}: L \to \mathfrak{P}(A)$] *is an order-embedding iff A is a base* [*respectively cobase*] *for L.*

roof Straightforward. □

If we combine propositions 8.6.2 and 8.6.3 we obtain:

roposition 8.6.4

$U_A^{\downarrow}: L \to \mathfrak{P}(A)$ [*respectively* $U_A^{\uparrow}: L \to \mathfrak{P}(A)$] *is a lattice-embedding iff A is an irreducible base* [*respectively a coirreducible cobase*] *for L.*

roof Immediate [propositions 8.6.2 and 8.6.3]. □

If $U_A^{\downarrow}$ is a lattice-embedding for some $A \subseteq L$, then it is one in the case when $A = \mathrm{Irr}(L)$. For this reason we restrict ourselves from now on to this special case and write simply $U^{\downarrow}$ instead of $U_{\mathrm{Irr}(L)}^{\downarrow}$. Dually, we write $U^{\uparrow}$ instead of $U_{\mathrm{Pt}(L)}^{\uparrow}$.

efinition

(1) L is said to be *spatial* if it has an irreducible base.
(2) L is *cospatial* if it has a coirreducible cobase.

roposition 8.6.5

$U^{\downarrow}: L \to \mathfrak{P}(\mathrm{Irr}(L))$ [*respectively* $U^{\uparrow}: L \to \mathfrak{P}(\mathrm{Pt}(L))$] *is a lattice-embedding iff L is spatial* [*respectively cospatial*].

roof Trivial [proposition 8.6.4]. □

In the case when L is Boolean there is a particularly simple criterion available to determine whether it is [co]spatial:

efinition

L is *atomic* if for every $a > \bot$ in L there is an atom $b \le a$ in L; L is *coatomic* if for every $a < \top$ in L there exists a coatom $b \ge a$ in L.

roposition 8.6.6

If L is Boolean then these four assertions are equivalent:
(*i*) *L is atomic;*
(*ii*) *L is coatomic;*
(*iii*) *L is spatial;*
(*iv*) *L is cospatial.*

(i)⇔(ii) Immediate [proposition 8.4.4].
(iii)⇔(iv) Ditto.

(i) ⇒ (iii) Suppose that L is atomic and that $a \not\leq b$. Then $a \sqcap b^* > \perp$ an so there exists an atom (hence irreducible) $c \leq a \sqcap b^*$. So $c \leq a$ and $c \leq b^*$ But if $c \leq b$ then $c = \perp$, contradicting the atomicity of c. So $c \not\leq b$. Thu L is spatial.
(iii) ⇒ (i) Immediate [proposition 8.6.1]. □

Example 2 In the distributive lattice $(\omega, |)$ the atoms are precisely the prime numbers, and s the lattice is atomic. The irreducibles, on the other hand, are the prime power Moreover, there are no coirreducibles (hence *a fortiori* no coatoms), and so th lattice is not coatomic. Hence the properties of atomicity and coatomicity are nc self-dual. This example also shows that the Boolean lattices are not characterize by the property that every coirreducible is a coatom (or, dually, by the propert that every irreducible is an atom).

Exercises

1. If L is Boolean, show that an element b of L is an atom iff b^* is a coatom.
2. L is said to be *disjunctive* if whenever $a \not\leq b$ in L there exists c in L such tha $b \sqcap c = \perp$ and $a \sqcap c > \perp$. (The dual concept is called a *conjunctive* lattice Show that every Boolean lattice is both conjunctive and disjunctive.

8.7 The representation of spatial lattices

If L is spatial, then the mapping $\mathrm{U}^{\downarrow} : L \to \mathfrak{P}(\mathrm{Irr}(L))$ is a lattice-embeddin and so its image $\mathrm{U}^{\downarrow}[L]$ is isomorphic to L. Now $\mathrm{Irr}(L)$ has an obviou partial order structure as a subcollection of L and we might hope t describe $\mathrm{U}^{\downarrow}[L]$ in terms of this structure. But unfortunately this is no possible in general, even if L is Boolean as well as spatial.

Example $\mathfrak{F}\mathfrak{C}(A)$ and $\mathfrak{P}(A)$ are both Boolean spatial lattices whose irreducible elements ar the singletons $\{a\}$ $(a \in A)$. Hence the partially ordered collections $\mathrm{Irr}(\mathfrak{F}\mathfrak{C}(A))$ an $\mathrm{Irr}(\mathfrak{P}(A))$ are isomorphic. But $\mathfrak{F}\mathfrak{C}(A)$ and $\mathfrak{P}(A)$ cannot be isomorphic if A i infinite.

Nevertheless, there are several special cases in which a solution to th problem is possible. We shall content ourselves with two of them.

Theorem 8.7.1 (Lindenbaum/Tarski)
These three assertions are equivalent:
(*i*) *L is a complete* [*co*]*atomic Boolean lattice*;
(*ii*) *the mapping* $\mathrm{U}^{\downarrow} : L \to \mathfrak{P}(\mathrm{Irr}(L))$ [*respectively* $\mathrm{U}^{\uparrow} : L \to \mathfrak{P}(\mathrm{Pt}(L))$] *is lattice isomorphism*;
(*iii*) *there exists a collection T such that L is isomorphic to* $\mathfrak{P}(T)$.

(i) ⇒ (ii) Suppose that L is complete and atomic. Then certainly L i spatial [proposition 8.6.6] and so $\mathrm{U}^{\downarrow} : L \to \mathfrak{P}(\mathrm{Irr}(L))$ is a lattice-embeddin

[proposition 8.6.5]. To show that it is surjective it is enough to show that $A = U^{\downarrow}(\sup A)$ for all $A \subseteq \mathrm{Irr}(L)$. But clearly $A \subseteq U^{\downarrow}(\sup A)$. So suppose that $A \subset U^{\downarrow}(\sup A)$, so that there exists $b \in U^{\downarrow}(\sup A) - A$. Now for all $x \in A$ we have $b \sqcap x = \bot$ since b and x are atoms [proposition 8.6.1], and so $x \leq b^*$. Consequently $\sup A \leq b^*$ and so $b = b \sqcap \sup A = \bot$. Contradiction.
(ii) ⇒ (iii) Obvious.
(iii) ⇒ (i) $\mathfrak{P}(T)$ is evidently complete and atomic; hence so is any lattice isomorphic to it. □

Theorem 8.7.2 (Birkhoff 1933)

These three assertions are equivalent:

(i) L is a finite distributive lattice;
(ii) L is finite and $U^{\downarrow} : L \to \mathfrak{C}^{\downarrow}(\mathrm{Irr}(L))$ *[respectively* $U^{\uparrow} : L \to \mathfrak{C}^{\uparrow}(\mathrm{Pt}(L))$*] is an isomorphism;*
(iii) there exists a finite partially ordered collection $(T, \leq)$ *such that L is isomorphic to* $\mathfrak{C}^{\downarrow}(T)$ *[respectively* $\mathfrak{C}^{\uparrow}(T)$*].*

(i) ⇒ (ii) Suppose that L is a finite distributive lattice. Then it is certainly partially well ordered and hence spatial. So $U^{\downarrow} : L \to \mathfrak{C}^{\downarrow}(\mathrm{Irr}(L))$ is a lattice-embedding [proposition 8.6.5]. To show that it is surjective it is enough to show that $A = U^{\downarrow}(\sup A)$ for every initial $A \subseteq \mathrm{Irr}(L)$. But clearly $A \subseteq U^{\downarrow}(\sup A)$. So suppose that $a \in U^{\downarrow}(\sup A)$. Then a is irreducible and $a \leq \sup A$. But A is finite and so $a \in \mathrm{Irr}(L) \cap \downarrow A = A$. Thus $U^{\downarrow}(\sup A) \subseteq A$ as required.
(ii) ⇒ (iii) Trivial.
(iii) ⇒ (i) It is easy to check that $\mathfrak{C}^{\downarrow}(T)$ is always a distributive lattice. □

Corollary 8.7.3

L is a finite Boolean lattice iff it is isomorphic to $\mathfrak{P}(T)$ *for some finite set T.*

Necessity This is a consequence of the Lindenbaum/Tarski theorem, since every finite distributive lattice is complete and atomic; equally it is a consequence of Birkhoff's theorem, since if L is Boolean the irreducibles of L coincide with the atoms [proposition 8.6.1] and are therefore totally unordered, so that $\mathfrak{C}^{\downarrow}(\mathrm{Irr}(L)) = \mathfrak{P}(\mathrm{Irr}(L))$.

Sufficiency Obvious. □

Remark As a consequence of corollary 8.7.3, note that for each $n \in \omega$ there is (up to isomorphism) exactly one Boolean lattice with 2^n elements and every finite Boolean lattice is one of these. On the other hand, there is a distributive (and hence Heyting) lattice of every finite cardinality $n \in \omega$

(**n** itself, for example) and it is easy to see that for $n \geq 4$ there is more than one.

Exercises

1. Show that every partially well ordered distributive lattice is complete, spatial and atomic.
2. Show that a complete lattice has a base consisting of atoms iff it is atomic and disjunctive.

9 The prime ideal property

In this chapter we shall show that (assuming the axiom of choice) every small distributive lattice L can be represented as (i.e. is isomorphic to) a sublattice of an appropriately chosen power set $\mathfrak{P}(\mathrm{spec}(L))$. This generalizes the results on the representation of spatial lattices which we obtained in §8.7. In the course of this endeavour we shall come across several lattice-theoretic assertions provably equivalent to the axiom of choice, and several (including the representation result itself) equivalent to a weaker principle called the 'prime ideal property'.

Throughout this chapter we use L and L′ to denote lattices, which will always be assumed to be distributive.

9.1 Ideals and filters

Definition

An *ideal* in L is an initial subcollection $\triangle$ of L such that for every finite $F \subseteq \triangle$ we have $\sup F \in \triangle$; dually, a *filter* of L is a final subcollection ∇ of L such that for every finite $F \subseteq \nabla$ we have $\inf F \in \nabla$.

It amounts to the same thing to say that $\triangle$ is an ideal of L if it satisfies:

$$x \leq y \in \triangle \Rightarrow x \in \triangle;$$

$$\bot \in \triangle;$$

$$x, y \in \triangle \Rightarrow x \sqcup y \in \triangle.$$

We leave it to the reader to state the corresponding properties that ∇ must satisfy to be a filter.

Example 1 $\{\bot\}$ is the smallest ideal in L and $\{\top\}$ is the smallest filter: they are said to be *trivial*.

Example 2 L is the largest ideal (and the largest filter) in L: it is said to be *improper*; all other ideals and filters are called *proper*. An ideal $\triangle$ of L is proper iff $\top \notin \triangle$; a filter ∇ in L is proper iff $\bot \notin \nabla$.

Example 3 Suppose that $f : L \to L'$ is a lattice homomorphism. If J is an ideal [respectively a filter] in L', then $f^{-1}[J]$ is an ideal [respectively a filter] in L. In particular,

$\Delta(f) = f^{-1}[\bot]$ is an ideal in L called the *kernel* of f; $\nabla(f) = f^{-1}[\top]$ is a filter in L called the *cokernel* of f.

Example 4 If $A \subseteq L$, then the smallest ideal in L containing A is

$$\downarrow[A] = \{x \in L : x \leq \sup F \text{ for some finite } F \subseteq A\}$$

and the smallest filter containing A is

$$\uparrow[A] = \{x \in L : x \geq \inf F \text{ for some finite } F \subseteq A\}.$$

Ideals [respectively filters] of the form $\downarrow a$ [respectively $\uparrow a$] with $a \in L$ are called *principal*.

Exercises

1. For each $a \in L$ we let

$$\Delta(a) = \{x \in L : a \sqcap x = \bot\}$$

$$\nabla(a) = \{x \in L : a \sqcup x = \top\}$$

Check that these are respectively an ideal and a filter in L.

2. (a) If ∇ is a filter in L, show that the smallest filter containing $\nabla \cup \{a\}$ is $\{y \in L : (\exists x \in \nabla)(y \geq x \sqcap a)\}$.
   (b) If $A \subseteq L$, show that $\uparrow A$ is a filter in L iff A is a filtered subcollection of L and that $\uparrow A$ is proper iff $\bot \notin A$. (Filtered subcollections of L are sometimes called *filter bases* in L. By analogy their duals, the directed subcollections of L, presumably ought to be, but for some reason are not, called 'ideal bases' in L.) Note that a filter is principal iff it has a base consisting of a single element.
   (c) If A, $B \subseteq L$, show that $\downarrow[A] \cap \uparrow[B] \neq \varnothing$ iff for some finite $F \subseteq A$ and $G \subseteq B$ we have $\sup F \geq \inf G$.
   (d) Show that the filter $\uparrow[A]$ is proper iff A has the *finite meet property*, i.e. $\inf F \neq \bot$ for all finite $F \subseteq A$.
3. Show that the collection $\mathrm{Idl}(L)$ of ideals in L is a frame with respect to inclusion and that there is a natural lattice-embedding $L \to \mathrm{Idl}(L)$ given by $x \mapsto \downarrow x$.

Proposition 9.1.1

If A, $B \subseteq L$, then

$$x[A, B]y \Leftrightarrow (\exists a \in \downarrow[A])(\exists b \in \uparrow[B])((x \sqcup a) \sqcap b = (y \sqcup a) \sqcap b).$$

Proof Let us write r to denote the relation on the right-hand side of the bi-implication. We shall prove first that r is a congruence. It is clear that r is reflexive and symmetric. To show that it is transitive, suppose that $x\,r\,y$ and $y\,r\,z$. So there exist $a, a' \in \downarrow[A]$ and $b, b' \in \uparrow[B]$ such that

$$(x \sqcup a) \sqcap b = (y \sqcup a) \sqcap b$$

and

$$(y \sqcup a') \sqcap b' = (z \sqcup a') \sqcap b'.$$

So

$$\begin{aligned}(x \sqcup (a \sqcup a')) \sqcap (b \sqcap b') &= ((x \sqcup a) \sqcap b \sqcap b') \sqcup (a' \sqcap b \sqcap b')\\ &= ((y \sqcup a) \sqcap b \sqcap b') \sqcup (a' \sqcap b \sqcap b')\\ &= (y \sqcup a \sqcup a') \sqcap b \sqcap b'\\ &= ((y \sqcup a') \sqcap b' \sqcap b) \sqcup (a \sqcap b \sqcap b')\\ &= ((z \sqcup a') \sqcap b' \sqcap b) \sqcup (a \sqcap b \sqcap b')\\ &= (z \sqcup (a \sqcup a')) \sqcap (b \sqcap b').\end{aligned}$$

Hence $x\ r\ z$ since $a \sqcup a' \in \downarrow[A]$ and $b \sqcap b' \in \uparrow[B]$. Now suppose just that $x\ r\ y$, so that there exist $a \in \downarrow[A]$ and $b \in \uparrow[B]$ such that

$$(x \sqcup a) \sqcap b = (y \sqcup a) \sqcap b.$$

Then

$$\begin{aligned}((x \sqcup z) \sqcup a) \sqcap b &= ((x \sqcup a) \sqcap b) \sqcup (z \sqcap b)\\ &= ((y \sqcup a) \sqcap b) \sqcup (z \sqcap b)\\ &= ((y \sqcup z) \sqcup a) \sqcap b,\end{aligned}$$

and so $x \sqcup z\ r\ y \sqcup z$.

Also

$$\begin{aligned}((x \sqcap z) \sqcup a) \sqcap b &= (x \sqcup a) \sqcap b \sqcap (z \sqcup a)\\ &= (y \sqcup a) \sqcap b \sqcap (z \sqcup a)\\ &= ((y \sqcap z) \sqcup a) \sqcap b,\end{aligned}$$

and so $x \sqcap z\ r\ y \sqcap z$.

This concludes the proof that r is a congruence. Notice now that if $a \in A$,

$$(a \sqcup a) \sqcap \top = a \sqcap \top = a = \bot \sqcup a = (\bot \sqcup a) \sqcap \top,$$

so that $a\ r\ \bot$. And if $b \in B$,

$$(b \sqcup \bot) \sqcap b = b \sqcap b = b = \top \sqcap b = (\top \sqcup \bot) \sqcap b,$$

so that $b\ r\ \top$. We have therefore shown that $[A, B] \subseteq r$, since $[A, B]$ is by definition the smallest congruence on L with these properties. To prove the converse, suppose that $x\ r\ y$, i.e. there exist $a \in \downarrow[A]$ and $b \in \uparrow[B]$ such that

$$(x \sqcup a) \sqcap b = (y \sqcup a) \sqcap b.$$

But then $a\,[A, B]\,\bot$ and $b\,[A, B]\,\top$. So

$$x = x \sqcup \bot\ [A, B]\ x \sqcup a = (x \sqcup a) \sqcap \top\ [A, B]\ (x \sqcup a) \sqcap b$$

and

$$y = y \sqcup \bot \, [A,B] \, y \sqcup a = (y \sqcup a) \sqcap \top \, [A,B] \, (y \sqcup a) \sqcap b.$$

Therefore $x \, [A,B] \, y$. This proves that $r \subseteq [A,B]$ as required. □

Corollary 9.1.2

Suppose that $A, B \subseteq L$.
(a) $[A,B]$ is improper iff $\downarrow[A] \cap \uparrow[B] \neq \varnothing$.
(b) $[A,B]$ is trivial iff $A \subseteq \{\bot\}$ and $B \subseteq \{\top\}$.

Proof (a) Suppose first that $\downarrow[A] \cap \uparrow[B] \neq \varnothing$. So there exists $c \in \downarrow[A] \cap \uparrow[B]$. Then $c \, [A,B] \, \bot$ and $c \, [A,B] \, \top$, so that $\bot \, [A,B] \, \top$, i.e. $[A,B]$ is improper. Suppose now that $[A,B]$ is improper. Then $\bot \, [A,B] \, \top$, i.e. there exist $a \in \downarrow[A]$ and $b \in \uparrow[B]$ such that $(\bot \sqcup a) \sqcap b = (\top \sqcup a) \sqcap b$ [proposition 9.1.1]. Hence $a \sqcap b = b$, i.e. $b \leq a$, and so $a, b \in \downarrow[A] \cap \uparrow[B]$.

(b) Suppose that $A \subseteq \{\bot\}$ and $B \subseteq \{\top\}$. So $\downarrow[A] = \{\bot\}$ and $\uparrow[B] = \{\top\}$. Hence

$$x \, [A,B] \, y \Leftrightarrow (x \sqcup \bot) \sqcap \top = (y \sqcup \bot) \sqcap \top \Leftrightarrow x = y$$

and so $[A,B]$ is trivial. Suppose finally that $[A,B]$ is trivial. Then the equivalence class of $\bot$ is $\{\bot\}$ and so $A \subseteq \{\bot\}$; similarly $B \subseteq \{\top\}$. □

Remark 1 In particular, if Δ is an ideal and ∇ is a filter in L then $[\Delta, \nabla]$ is proper iff Δ and ∇ are disjoint, trivial iff Δ and ∇ are both trivial.

Remark 2 If $f : L \to L'$ is a lattice homomorphism then the congruence $[\Delta(f), \nabla(f)]$ is improper iff L' has only one element; and if f is injective then the congruence is trivial. However, the converse of this last assertion need not hold; Fig. 8 illustrates a counterexample. This is a case where the analogy between lattice theory and ring theory breaks down: if the kernel of a ring homomorphism is trivial, the mapping must be injective.

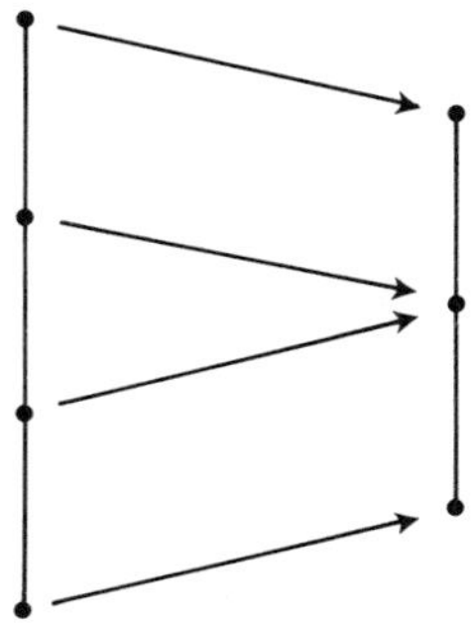

Figure 8

Exercise 4. If L is Boolean and $f : L \to L'$ is a lattice homomorphism such that $[\Delta(f), \nabla(f)]$ is trivial, show that $f : L \to L'$ is injective.

9.2 The spectrum

In §8.6 we showed that if L is spatial [respectively cospatial] then it can be embedded in $\mathfrak{P}(\mathrm{Irr}(L))$ [respectively $\mathfrak{P}(\mathrm{Pt}(L))$]. But many otherwise well-behaved lattices are neither spatial nor cospatial. If $L = \mathrm{Bool}(\omega)$, for instance, then $\mathrm{Irr}(L) = \mathrm{Pt}(L) = \varnothing$. If we are to obtain a general representation theorem, it seems that we should look for a way of associating with L a collection which is in general larger than either $\mathrm{Pt}(L)$ or $\mathrm{Irr}(L)$. Here is the definition we propose:

Definition

A *character* of L is a lattice homomorphism $L \to \mathbf{2}$. The collection of all characters of L is called the *spectrum* of L and denoted $\mathrm{spec}(L)$.

Since a character takes only two values, it is determined entirely by its kernel [respectively cokernel]. Ideals [respectively filters] which arise in this way will play an important role in what follows—in particular, they often prove easier to visualize and to manipulate than the corresponding characters—and we single them out now under the name 'prime'.

Definition

An ideal Δ of L is said to be *prime* if for every finite $F \subseteq L$ such that $\inf F \in \Delta$ there exists $x \in F$ such that $x \in \Delta$; dually, a filter ∇ of L is said to be *prime* if for every finite $F \subseteq L$ such that $\sup F \in \nabla$ there exists $x \in F$ such that $x \in \nabla$.

It amounts to the same thing to say that an ideal Δ of L is prime iff it satisfies:

$$\top \notin \Delta;$$

$$x \sqcap y \in \Delta \Rightarrow (x \in \Delta \text{ or } y \in \Delta).$$

Once again we leave it to the reader to formulate the dual assertion.

Proposition 9.2.1

(*a*) *There is a definite bijective mapping from* $\mathrm{spec}(L)$ *onto the collection of prime filters of* L *given by* $f \mapsto \nabla(f)$; *its inverse is given by* $\nabla \mapsto \mathrm{c}_\nabla$ (*where* c_∇ *is the characteristic function of* ∇).

(*b*) *There is a definite bijective mapping from* $\mathrm{spec}(L)$ *onto the collection of prime ideals of* L *given by* $f \mapsto \Delta(f)$; *its inverse is given by* $\Delta \mapsto \mathrm{c}_{L-\Delta}$.

Proof Straightforward. □

1. Suppose that $f: L \to L'$ is a lattice homomorphism and L' is non-trivial.
   (a) Show that if J is a prime ideal [respectively prime filter] in L' then $f^{-1}[J]$ is a prime ideal [respectively filter] in L.
   (b) Show that $\triangle(f)$ [respectively $\nabla(f)$] is a prime ideal [respectively prime filter] in L iff $\text{im}[f] = \{\bot, \top\}$.

2. (a) Show that the filter $\uparrow a$ is prime iff a is irreducible.
   (b) If ∇ is a maximal proper filter in L, show that these three assertions are equivalent:
      (i) $\nabla = \uparrow a$ for some atom $a \in L$;
      (ii) ∇ is principal;
      (iii) $\inf \nabla \neq \bot$.
   (c) If L is Boolean, show that a maximal proper filter in L is principal iff it has a countable base.
   (d) Show that the Boolean lattice $\mathfrak{FC}(\omega)$ has exactly one maximal proper filter which is not principal. How many does the product Boolean lattice $\mathfrak{FC}(\omega) \times \mathbf{2}$ have?

3. If L and L' are frames, then a *frame homomorphism* is a lattice homomorphism $f: L \to L'$ such that $f(\sup B) = \sup f[B]$ for all $B \subseteq L$. Show that if $f: L \to L'$ is a frame homomorphism then $\triangle(f)$ is a principal ideal in L.

Proposition 9.2.2

There is a one-to-one correspondence between the collection of prime ideals of L and the collection of prime filters given by $\triangle \mapsto L \to \triangle$; its inverse is given by $\nabla \mapsto L - \nabla$.

Proof   Trivial.   □

Let us return now to a remark we made when we introduced the notion of a 'spectrum'. We said then that we wanted $\text{spec}(L)$ to be in general larger than either $\text{Pt}(L)$ or $\text{Irr}(L)$. We shall now substantiate this.

Definition

(1) A character f of L is said to be *irreducible* if there exists $a \in L$ such that $f(x) = \top \Leftrightarrow x \geq a$; the collection of irreducible characters of L is denoted $\text{irr}(L)$.
(2) A character f of L is called a *point* if there exists $a \in L$ such that $f(x) = \bot \Leftrightarrow x \leq a$; the collection of all points of L is denoted $\text{pt}(L)$.

Recall that the collection of all irreducible [respectively coirreducible] elements of L is denoted $\text{Irr}(L)$ [respectively $\text{Pt}(L)$]. The similarity in the notation is intentional: we shall now show that there is a definite bijective mapping between the two collections.

Proposition 9.2.3

(a) There is a definite bijective mapping $f^{\downarrow} : \mathrm{Irr}(L) \to \mathrm{irr}(L)$ *given by*

$$f^{\downarrow}(a)(x) = \begin{cases} \top & \text{if } x \geq a \\ \bot & \text{otherwise.} \end{cases}$$

(b) There is a definite bijective mapping $f^{\uparrow} : \mathrm{Pt}(L) \to \mathrm{pt}(L)$ *given by*

$$f^{\uparrow}(a)(x) = \begin{cases} \bot & \text{if } x \leq a \\ \top & \text{otherwise.} \end{cases}$$

Proof Straightforward. □

Theorem 9.2.4

Suppose that ∇ *is a filter in L and* $\mathscr{A}$ *is the collection of all ideals in L which are disjoint from* ∇.
(a) $\mathscr{A}$ *has finite character.*
(b) Every maximal element of $\mathscr{A}$ *is a prime ideal in L.*

Proof (a) Straightforward exercise.
(b) Suppose that $\triangle$ is a maximal element of $\mathscr{A}$. Since $\triangle$ and ∇ are disjoint, $L/[\triangle, \nabla]$ is non-trivial [corollary 9.1.2(a)]. If $\mathrm{p} : L \to L/[\triangle, \nabla]$ is the canonical homomorphism, then $\triangle \subseteq \triangle(\mathrm{p})$ and $\nabla \subseteq \nabla(\mathrm{p})$. Moreover $L/[\triangle, \nabla]$ contains exactly two elements: otherwise it would contain a non-trivial proper ideal J, and then $\mathrm{p}^{-1}[J]$ would be an ideal in L disjoint from ∇ and strictly containing $\triangle$, which contradicts the maximality of $\triangle$ in $\mathscr{A}$. It follows that $\triangle(\mathrm{p})$ is prime. But $\triangle = \triangle(\mathrm{p})$ by the maximality of $\triangle$. Thus $\triangle$ is prime, which is what we wanted to prove. □

Corollary 9.2.5

Every maximal proper ideal in L is prime.

Proof Let $\nabla = \{\top\}$ and apply theorem 9.2.4(b). □

Proposition 9.2.6

If L is Boolean and $\triangle$ *is an ideal of L then these three assertions are equivalent:*
(i) $\triangle$ *is prime;*
(ii) for each $a \in L$ *exactly one of* a, a^* *is an element of* $\triangle$;
(iii) $\triangle$ *is a maximal proper ideal of L.*

(i) $\Rightarrow$ (ii) $a \sqcap a^* = \bot \in \triangle$. So either $a \in \triangle$ or $a^* \in \triangle$. But $a \sqcup a^* = \top \notin \triangle$ and so we cannot have both $a \in \Delta$ and $a^* \in \triangle$.

(ii) $\Rightarrow$ (iii) $\bot \in \Delta$ and hence $\top = \bot^* \notin \Delta$, so that Δ is proper. Moreover, if I is a proper ideal of L which strictly contains Δ and $a \in I - \Delta$, then by hypothesis $a^* \in \Delta \subseteq I$ and so $\top = a \sqcup a^* \in I$, contradicting the assumption that I is proper. This shows that Δ is maximal.
(iii) $\Rightarrow$ (i) Immediate [corollary 9.2.5]. □

Of course, all these results have duals, which we leave to the reader to formulate.

Example 1 If L is partially well ordered (in particular, if L is finite), then $\text{irr}(L) = \text{spec}(L)$, and so $f^{\downarrow}$ is a definite one-to-one correspondence between $\text{Irr}(L)$ and $\text{spec}(L)$.

Example 2 Dually, if L is Noetherian, then $f^{\uparrow}$ provides a definite one-to-one correspondence between $\text{Pt}(L)$ and $\text{spec}(L)$.

Example 3 If A is a collection then for each $a \in A$ there is a character of $\mathfrak{FC}(A)$ given by

$$f_a(B) = \begin{cases} \top & \text{if } a \in B \\ \bot & \text{otherwise.} \end{cases}$$

The corresponding prime filter on $\mathfrak{FC}(A)$ is the principal filter generated by the singleton $\{a\}$. If A is not finite there is also a character f_∞ given by

$$f_\infty(B) = \begin{cases} \top & \text{if } B \text{ is cofinite} \\ \bot & \text{if } B \text{ is finite.} \end{cases}$$

The corresponding prime filter is $\{B \subseteq A : B \text{ is cofinite}\}$. It is a straightforward exercise to check that there are no other characters (and hence no other prime filters). Let us adopt the convention that A^+ denotes A itself if A is finite and $A \cup \{\infty\}$ if not, where ∞ is some definite object not in A (such as A itself). Then $x \mapsto f_x$ is a definite one-to-one correspondence between A^+ and $\text{spec}(\mathfrak{FC}(A))$.

Example 4 Each mapping $f : A \to \mathbf{2}$ gives rise to a character $\bar{f} : \text{Bool}(A) \to \mathbf{2}$; it is easy to check that the function $f \mapsto \bar{f}$ is a definite one-to-one correspondence between ${}^A\mathbf{2}$ and $\text{Spec}(\text{Bool}(A))$.

Exercises

4. Prove theorem 9.2.4(a).
5. Suppose that L is complete.
   (a) A filter ∇ in L is said to be *completely prime* if for every $B \subseteq L$ it is the case that $\sup B \in \nabla \Rightarrow B \cap \nabla \neq \varnothing$. Show that this is the case iff its complement is a principal prime ideal in L.
   (b) For any mapping $f : L \to \mathbf{2}$ show that these four assertions are equivalent:
      (i) $f : L \to \mathbf{2}$ is a frame homomorphism;
      (ii) $\Delta(f)$ is a principal prime ideal in L;
      (iii) $\nabla(f)$ is a completely prime filter in L.
      (iv) f is a point.
6. Suppose that L is a totally ordered collection with a greatest and a least element.

(a) Show that every proper ideal of L is prime.
(b) If L has more than two elements, show that it has a prime ideal which is not maximal.

7. If L is partially well ordered, show that every filter in L is principal.

9.3 The prime ideal property

We are now ready to embark on another attempt to embed every distributive lattice in a power lattice. In our first, only partially successful, attempt (in §8.6) we used the mappings $\mathrm{U}^{\downarrow} : L \to \mathfrak{P}(\mathrm{Irr}(L))$ and $\mathrm{U}^{\uparrow} : L \to \mathfrak{P}(\mathrm{Pt}(L))$ given by

$$\mathrm{U}^{\downarrow}(x) = \mathrm{Irr}(L) \cap \downarrow x$$
$$\mathrm{U}^{\uparrow}(x) = \mathrm{Pt}(L) - \uparrow x.$$

This time we intend to embed L in the potentially larger power lattice $\mathfrak{P}(\mathrm{spec}(L))$. To get a clue as to how to proceed let us observe that if we identify $\mathrm{Irr}(L)$ with $\mathrm{irr}(L)$ and $\mathrm{Pt}(L)$ with $\mathrm{pt}(L)$ by means of $f^{\downarrow}$ and $f^{\uparrow}$ respectively then $U^{\downarrow}$ and $U^{\uparrow}$ can be thought of as mappings $L \to \mathfrak{P}(\mathrm{irr}(L))$ and $L \to \mathfrak{P}(\mathrm{pt}(L))$ respectively. Put more formally (i.e. dodging the precise meaning of 'identify' here), we define mappings $\mathrm{V}^{\downarrow} : L \to \mathfrak{P}(\mathrm{irr}(L))$ and $\mathrm{V}^{\uparrow} : L \to \mathfrak{P}(\mathrm{pt}(L))$ by

$$\mathrm{V}^{\downarrow}(x) = f^{\downarrow}[\mathrm{U}^{\downarrow}(x)]$$
$$\mathrm{V}^{\uparrow}(x) = f^{\uparrow}[\mathrm{U}^{\uparrow}(x)].$$

Then a straightforward calculation shows that

$$\mathrm{V}^{\downarrow}(x) = \{f \in \mathrm{irr}(L) : f(x) = \top\}$$
$$\mathrm{V}^{\uparrow}(x) = \{f \in \mathrm{pt}(L) : f(x) = \top\}.$$

The conclusion should be obvious: we ought to try defining a mapping $L \to \mathfrak{P}(\mathrm{spec}(L))$ as follows.

Definition

We define $\mathrm{V} : L \to \mathfrak{P}(\mathrm{spec}(L))$ by

$$\mathrm{V}(x) = \{f \in \mathrm{spec}(L) : f(x) = \top\}.$$

Lemma 9.3.1

$\mathrm{V} : L \to \mathfrak{P}(\mathrm{spec}(L))$ *is a lattice homomorphism.*

Proof Trivial exercise. □

Theorem 9.3.2

These three assertions are equivalent:

(*i*) *The axiom of choice.*

(*ii*) *If L is small, I is an ideal in L and F is a filter in L such that $I \cap F = \varnothing$, then there exists an ideal Δ in L which is maximal among those containing I and disjoint from F* (Stone 1936).

(*iii*) *Every non-trivial small distributive lattice contains a maximal proper ideal* (The maximal ideal property).

(i) ⇒ (ii) The set of all ideals of L disjoint from F has finite character [theorem 9.2.4(a)] and hence each element of it is contained in a maximal element [Teichmüller/Tukey property].
(ii) ⇒ (iii) Apply (ii) with $I = \{\bot\}$ and $F = \{\top\}$.
(iii) ⇒ (i) Let $\mathscr{A}$ be a set of non-empty sets: it will be enough [lemma 7.1.4(b)] to show that choice($\mathscr{A}$) has a maximal element. So let L be the sublattice of $\mathfrak{P}$(choice($\mathscr{A}$)) generated by $\{\uparrow f : f \in \text{choice}(\mathscr{A})\}$. Now we can assume that at least one member of $\mathscr{A}$ has more than one element since otherwise the problem is trivial. Consequently L is non-trivial, so by the dual of (iii) there is a maximal proper filter ∇ in L. Let $\mathscr{B} = \{f \in \text{choice}(\mathscr{A}) : \uparrow f \in \nabla\}$. Now if $f, g \in \mathscr{B}$ then $\uparrow f \cap \uparrow g \neq \varnothing$ and so $f \cup g$ is a function: it follows [lemma 2.2.1] that $h = \bigcup \mathscr{B}$ is a function and $h \in \text{choice}(\mathscr{A})$. If $C \in \nabla$ then $C = \uparrow f_0 \cup \cdots \cup \uparrow f_{n-1}$ for some $f_0, \ldots, f_{n-1} \in \mathscr{B}$; since ∇ is prime, there exists r such that $\uparrow f_r \in \nabla$, i.e. $f_r \in \mathscr{B}$, and so $\uparrow h \subseteq \uparrow f_r \subseteq C$. Thus $\uparrow h \subseteq \bigcap \nabla$. Consequently ∇ is the principal filter in L generated by $\uparrow h$ and so $\uparrow h$ is an atom of L (cf. exercise 2(a) of §9.2). Therefore h is a maximal element of choice($\mathscr{A}$), which is what we wanted. □

The properties stated in parts (ii) and (iii) of theorem 9.3.2 hold for countable lattices without the axiom of choice.

Proposition 9.3.3

If L is countable, I is an ideal in L and F is a filter in L such that $I \cap F = \varnothing$, then there exists an ideal Δ in L which is maximal among those disjoint from F and containing I.

Proof Immediate [theorem 9.2.4(a) and proposition 7.1.6(d)]. □

We shall soon see that if we assume the axiom of choice then V is an embedding. However, the way in which we intend to demonstrate this is somewhat indirect: we shall prove first that the assertion that V is injective is equivalent to a rather weaker axiom called the prime ideal property.

Definition

The [*Boolean*] *prime ideal property* is the assertion that every non-trivial small [Boolean] distributive lattice has a prime ideal.

Proposition 9.3.4

The axiom of choice implies the prime ideal property.

Proof Trivial [theorem 9.3.2 and corollary 9.2.5]. □

Theorem 9.3.5

These five assertions are equivalent:

(*i*) *The prime ideal property.*

(*ii*) *If L is small, I is an ideal of L, and F is a filter of L such that $I \cap F = \varnothing$, then there exists a prime ideal in L which contains I and is disjoint from F.*

(*iii*) *If L is small, $a, b \in L$ and $b \nleq a$, then there exists $f \in \mathrm{spec}(L)$ such that $f(a) = \bot$ and $f(b) = \top$* (separation property).

(*iv*) *If L is small, then the canonical mapping $\mathrm{V} : L \to \mathfrak{P}(\mathrm{spec}(L))$ is a lattice-embedding* (Birkhoff 1933).

(*v*) *If L is small, then it can be embedded as a lattice in $\mathfrak{P}(T)$ for some set T.*

(i) ⇒ (ii) Since $I \cap F \neq \varnothing$, the congruence $[I, F]$ is proper [corollary 9.1.2]; so the distributive quotient lattice $L/[I, F]$ is non-trivial and therefore by hypothesis has a prime ideal $\triangle$. So if $\mathrm{p} : L \to L/[I, F]$ is the canonical quotient mapping, then $\mathrm{p}^{-1}[\triangle]$ is a prime ideal in L which contains I and is disjoint from F.

(ii) ⇒ (iii) $\downarrow a$ is an ideal of L and $\uparrow b$ is a filter; moreover, they are disjoint because $b \nleq a$. So by hypothesis there exists a prime ideal containing $\downarrow a$ and disjoint from $\uparrow b$. If f is the associated character then evidently $f(a) = \bot$ and $f(b) = \top$.

(iii) ⇒ (iv) We have already remarked that $\mathrm{V} : L \to \mathfrak{P}(\mathrm{spec}(L))$ is a lattice homomorphism. So we only need to show that it is injective. Now if $a \neq b$ then either $b \nleq a$ or $a \nleq b$: so there exists $f \in \mathrm{spec}(L)$ such that $f(a) \neq f(b)$, and consequently $\mathrm{V}(a) \neq \mathrm{V}(b)$.

(iv) ⇒ (v) Trivial.

(v) ⇒ (i) If L is non-trivial and $f : L \to \mathfrak{P}(T)$ is a lattice-embedding, then T has an element $t \in T$; if we let $\triangle = \{a \in L : t \notin f(a)\}$, it is trivial to check that $\triangle$ is a prime ideal of L. □

Exercises

1. Let M be a (not necessarily distributive) lattice. Assuming the prime ideal property, show that M is distributive iff every proper ideal of M is the intersection of the prime ideals containing it.
2. Show that $\mathrm{V}[A] = \mathrm{V}[\downarrow[A]]$ for all $A \subseteq L$.
3. Let L be a frame.

(a) Show that $V^{\uparrow} : L \to \mathfrak{P}(\mathrm{pt}(L))$ is a frame homomorphism.
(b) Show that $U^{\uparrow}(x) = \{\sup \Delta(f) : f \in V(x)\}$ for all $x \in L$.
(c) If $a \in L$, show that these two assertions are equivalent:
  (i) If $A \subseteq L$ and $\sup A = a$ then there exists a finite set $F \subseteq A$ such that $\sup F = a$.
  (ii) If I is an ideal in L and $\sup I \geq a$ then $a \in I$.
  The element $a \in L$ is said to be *compact* if it satisfies these conditions.
(d) Show that the supremum of a finite collection of compact elements of L is compact.
(e) L itself is said to be *compact* if $\top$ is a compact element of L. If L is compact, show that every proper ideal in L is contained in a principal proper ideal.
(f) Show that an element of $\mathfrak{P}(A)$ is compact iff it is a finite subset of A.

4. A frame M is said to be *coherent* if the set $\mathfrak{K}(M)$ of compact elements of M is both a sublattice of M and a base for M.
  (a) Show that if L is a distributive lattice then the collection $\mathrm{Idl}(L)$ of all the ideals of L is a coherent frame with respect to inclusion, that the coirreducible and the compact elements of $\mathrm{Idl}(L)$ are respectively the prime and the principal ideals of L, and that the mapping $L \to \mathfrak{K}(\mathrm{Idl}(L))$ given by $a \mapsto \downarrow a$ is an isomorphism.
  (b) Show that if M is a coherent frame then it is isomorphic to $\mathrm{Idl}(\mathfrak{K}(M))$.
5. (a) Assuming the axiom of choice, show that every non-trivial compact frame has at least one point.
  (b) (Banaschewski 1985) Prove the same result assuming only the prime ideal property.

9.4 The partial order of the spectrum

If we assume the prime ideal property then Birkhoff's representation theorem tells us that L can be regarded in a natural way as a sublattice of $\mathfrak{P}(\mathrm{spec}(L))$. But which one? It is easy to see that $\mathrm{spec}(L)$, when thought of simply as a set, does not have encoded in it enough information to answer this question.

Example 1 Consider the two distributive lattices in Fig. 9. The spectrum of each lattice has two elements, but they are clearly not isomorphic.

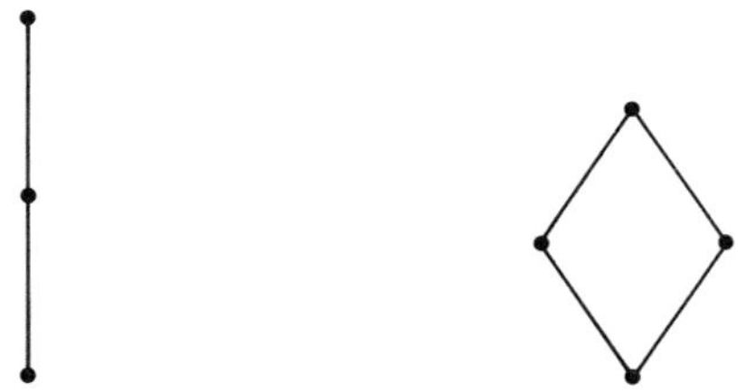

(a) (b)

Figure 9

We must therefore give spec(L) some more structure if we are to retrieve a complete description of L from it. When we addressed the same problem in relation to Irr(L) and Pt(L) in §8.7, we met with some success by considering the partial orderings on these collections which they inherit from L by reason of being subcollections of L. We can evidently transfer these partial orderings to irr(L) and pt(L) by means of the identification mappings $f^{\downarrow}$ and $f^{\uparrow}$. More formally, if $a, b \in \mathrm{Irr}(L)$ and $a \leq b$ then

$$f^{\downarrow}(b)(x) = \top \Leftrightarrow x \geq b \Rightarrow x \geq a \Leftrightarrow f^{\downarrow}(a)(x) = \top,$$

so that $f^{\downarrow}(a)(x) \geq f^{\downarrow}(b)(x)$ for all $x \in L$. And if, conversely, $f^{\downarrow}(a)(x) \geq f^{\downarrow}(b)(x)$ for all $x \in L$, then in particular

$$f^{\downarrow}(a)(b) \geq f^{\downarrow}(b)(b) = \top$$

and so $a \leq b$. In other words, the partial ordering on irr(L) is given by

$$f \leq g \Leftrightarrow f(x) \geq g(x) \quad \text{for all } x \in L.$$

Similar trivial checking shows that the partial ordering on pt(L) is defined by the same equivalence. These observations lead us to the following definition.

Definition

If $f, g \in \mathrm{spec}(L)$ then we define

$$f \leq g \Leftrightarrow f(x) \geq g(x) \quad \text{for all } x \in L.$$

Warning Notice that this is the *opposite* of the partial ordering one would naively expect.

Example 2 If L is Noetherian, then spec(L) is order isomorphic to Pt(L).

Example 3 If L is finite, then both L and L^{op} are partially well ordered and so spec(L) is order isomorphic to both Pt(L) and Irr(L).

Example 4 If L is Boolean, then spec(L) is totally unordered by $\leq$ (see theorem 9.4.2 below) and is therefore determined (up to isomorphism) by its cardinality.

Lemma 9.4.1

Every maximal proper ideal in L is a maximal prime ideal. If we assume the prime ideal property, then the converse is also true (provided that L is small).

Necessity If Δ is a maximal proper ideal in L, then it is prime [corollary 9.2.5]; indeed it is maximally prime since every prime ideal is proper.

Sufficiency If Δ is a maximal prime ideal then it is certainly proper; if it were not a maximal proper ideal, it would be strictly contained in a proper ideal, which in turn would be contained in a prime ideal by the prime ideal property, contradicting the maximality of Δ. □

Remark It follows at once from this lemma that the natural one-to-one correspondence between spec(L) and the set of prime ideals in L restricts (assuming the prime ideal property) to a correspondence between the set max(L) of maximal elements of spec(L) and the set of maximal proper ideals in L; the set min(L) of minimal elements of spec(L) corresponds dually to the set of maximal proper filters in L.

We shall now show that the partially ordered set (spec(L), $\leq$) does at any rate retain enough information to decide whether L is Boolean or not.

Theorem 9.4.2 (Nachbin 1947, 1949)

If we assume the prime ideal property, then these three assertions are equivalent (provided that L is small):

(i) L is Boolean;

(ii) every prime ideal of L is a maximal proper ideal;

(iii) (spec(L), $\leq$) *is totally unordered.*

(i) ⇒ (ii) Proposition 9.2.6.

(ii) ⇔ (iii) Immediate [lemma 9.4.1].

(ii) ⇒ (i) Suppose that $a \in L$ is not complemented and let $\nabla = \uparrow[\nabla(a) \cup \{a\}]$. Now

$$\begin{aligned} \bot \in \nabla &\Rightarrow (\exists b \in \nabla(a))(a \sqcap b = \bot) \\ &\Rightarrow (\exists b \in L)(a \sqcap b = \bot \text{ and } a \sqcup b = \top) \\ &\Rightarrow a \text{ is complemented} \\ &\Rightarrow \text{contradiction.} \end{aligned}$$

So $\bot \notin \nabla$, i.e. the ideal $\{\bot\}$ and the filter ∇ are disjoint. So since we are assuming the prime ideal property there exists a prime ideal Δ in L disjoint from ∇. Let $I = \downarrow[\Delta \cup \{a\}]$. Then

$$\begin{aligned} \top \in I &\Rightarrow (\exists b \in \Delta)(\top = a \sqcup b) \\ &\Rightarrow (\exists b \in \Delta)(b \in \nabla(a) \subseteq \nabla) \\ &\Rightarrow \Delta \cap \nabla \neq \varnothing \\ &\Rightarrow \text{contradiction.} \end{aligned}$$

So $\top \notin I$, i.e. I is a proper ideal of L containing Δ. Thus Δ is not a maximal proper ideal of L despite being prime. □

Exercises

1. If L is partially well ordered and every irreducible element of L is an atom, show that L is Boolean.

2. (a) Show that the spectra of $\mathfrak{F}\mathfrak{C}(\omega)$ and $\mathfrak{F}\mathfrak{C}(\omega) \times \mathbf{2}$ are both countably infinite.
   (b) Deduce that the spectra are order isomorphic, although the lattices themselves are not isomorphic (see exercise 2(d) of §9.2).

Unfortunately, (spec(L), $\leq$) is not sufficient to determine L completely, as the above exercise shows. We shall have to wait a little longer before we can define a structure on spec(L) which determines L in all cases.

9.5 Topological spaces

You are probably familiar with the elementary facts about topological spaces. As a brief reminder:

Definition

A *topology* on X is a collection $\mathscr{T}$ of subcollections of X such that:

$$\text{if } \mathscr{A} \text{ is a subcollection of } \mathscr{T} \text{ then } \bigcup \mathscr{A} \in \mathscr{T};$$

$$\text{if } \mathscr{F} \text{ is a finite subcollection of } \mathscr{T} \text{ then } \bigcap \mathscr{F} \in \mathscr{T}.$$

The pair $(X, \mathscr{T})$ is then called a *topological space* (or sometimes just a *space*).

The elements of $\mathscr{T}$ are said to be *open* in X; their relative complements in X are said to be *closed* in X. The definition amounts to the requirements that:
if I is non-empty and U_i is open for all $i \in I$ then $\bigcup_{i \in I} U_i$ is open in X;
if U and V are open in X then $U \cap V$ is open in X;
$\varnothing$ and X itself are both open in X.
The last property follows from the definition because $\bigcup \varnothing = \varnothing$ and (by convention) $\bigcap \varnothing = X$.

If $A \subseteq X$ then the largest open set contained in A is called the *interior* of A and denoted Int(A); the smallest closed set containing A is called the *closure* of A and denoted Cl(A).

If $\mathscr{T}$ is the smallest topology on X containing a given collection $\mathscr{B}$ of subcollections of X, then we say that $\mathscr{T}$ is *generated* by $\mathscr{B}$ or that $\mathscr{B}$ is a *subbasis* for $\mathscr{T}$. If in addition every element of $\mathscr{T}$ is the union of a subcollection of $\mathscr{B}$ then $\mathscr{B}$ is said to be a *basis* for $\mathscr{T}$. It is easy to verify that this is the case iff for every finite $\mathscr{F} \subseteq \mathscr{B}$ we have

$$x \in \bigcap \mathscr{F} \Rightarrow (\exists B \in \mathscr{B})(x \in B \subseteq \bigcap \mathscr{F}).$$

Equivalently, $\mathscr{B}$ is a basis iff:

$\mathscr{B}$ is a covering of X, i.e. $X = \bigcup \mathscr{B}$;

if $B_1, B_2 \in \mathscr{B}$ and $x \in B_1 \cap B_2$ then there exists $B \in \mathscr{B}$ such that $x \in B \subseteq B_1 \cap B_2$.

Definition

A topological space $(X, \mathscr{T})$ (or a topology $\mathscr{T}$) is said to be *Hausdorff* if for any two distinct points $x, y \in X$ there exist disjoint open sets U and V in X such that $x \in U$ and $y \in V$.

Lemma 9.5.1

If $(X, \mathscr{T})$ is a topological space and $\mathscr{B}$ is a basis for the topology of X then these three assertions are equivalent:

(i) *Every covering of X by open collections has a finite subcollection which is a covering of X.*

(ii) *Every covering of X by elements of $\mathscr{B}$ has a finite subcollection which is a covering of X.*

(iii) *If $\mathscr{A}$ is a collection of closed subcollections of X which has the finite intersection property (i.e. every finite subcollection of $\mathscr{A}$ has non-empty intersection) then $\mathscr{A}$ has non-empty intersection.*

Proof Exercise. □

Definition

A topological space $(X, \mathscr{T})$ (or a topology $\mathscr{T}$) is said to be *compact* if it satisfies the equivalent conditions of lemma 9.5.1.

Natural examples of topologies abound in all parts of mathematics. Most of those which arise in elementary geometry and analysis are metric topologies: in other words, they have as a subbasis the open balls of some distance-measuring function (metric) appropriate to the study being undertaken. However, the topologies which occur in algebra are rarely of this kind.

Example 1 Suppose that $(A, \leq)$ is partially ordered. A topology on A is called *upper* if $\mathrm{Cl}(a) = \downarrow a$ for all $a \in A$. The final subcollections of A form an upper topology on A; in fact, it is the largest one and is called the *strong upper topology* on A. The smallest upper topology on A is the one generated by the collections $A - \downarrow a$ for $a \in A$; it is called the *weak upper topology* on A. The corresponding topologies on A^{op} are called *lower* topologies.

Example 2 The smallest topology on the partially ordered collection $(A, \leq)$ which contains both the weak upper topology and the weak lower topology is called the *order topology*. The case when A is totally ordered by $\leq$ is the most important one: in this case the order topology has a basis consisting of the open intervals of A

When we refer to an ordered collection as a topological space without further qualification it is generally this topology we shall be referring to. For instance, the order topology on $\mathbf{R}$ is just the familiar metric topology. The space $\boldsymbol{\alpha}$ (where α is an ordinal) is a fruitful source of examples in topology: it is always Hausdorff but is compact iff α is a successor ordinal.

Example 3 The subcollections A of X such that $X - A$ is finite form (together with $\varnothing$) a topology on X called the *cofinite* topology. It is always compact but is Hausdorff iff X is finite.

Exercises

1. Prove lemma 9.5.1.
2. Let us call a point $b \in X$ *isolated* if $\{b\}$ is open in X and *limit* otherwise. Show that an ordinal $\beta \in \boldsymbol{\alpha}$ is a limit point of $\boldsymbol{\alpha}$ iff β is a limit ordinal.
3. (a) If X is a topological space, then we denote by $\mathfrak{O}(X)$ the collection of open subcollections of X partially ordered by inclusion. Show that $\mathfrak{O}(X)$ is a spatial frame. [The supremum of a collection $\mathscr{A} \subseteq \mathfrak{O}(X)$ is $\bigcup \mathscr{A}$ but the infimum is $\operatorname{Int}(\bigcap \mathscr{A})$.]
   (b) Show that the lattice $\mathfrak{A}(X)$ of closed subcollections of X partially ordered by inclusion is isomorphic to $\mathfrak{O}(X)^{\mathrm{op}}$.
   (c) Show that X is compact iff $\mathfrak{O}(X)$ is a compact frame.
   (d) Show that no corresponding characterization of the Hausdorffness of X in terms of a property of the frame $\mathfrak{O}(X)$ is possible.
   (e) Show that if X is Hausdorff then every irreducible element of $\mathfrak{A}(X)$ is an atom.
   (f) Show that $\mathfrak{O}(X)$ [respectively $\mathfrak{A}(X)$] is Boolean iff every open subcollection of X is closed.
   (g) Deduce that the closed subsets of $\mathbf{R}$ form a non-Boolean distributive lattice in which every irreducible element is an atom.
4. A space X is said to be *connected* if it does not have a non-empty proper subcollection which is both closed and open. Prove that X is non-empty and connected iff $\mathfrak{B}\mathfrak{O}(X)$ has exactly two elements.
5. A space X is said to be *irreducible* if every finite intersection of non-empty open subcollections of X is non-empty.
   (a) Show that X is irreducible iff $\mathfrak{R}\mathfrak{O}(X)$ has exactly two elements.
   (b) Show that every irreducible Hausdorff space has exactly one element.
   (c) Give an example of an infinite irreducible space X in which $\{x\}$ is closed for all $x \in X$. [$\boldsymbol{\omega}$ with the cofinite topology.]

9.6 Continuous mappings

Lemma 9.6.1

If $(X, \mathscr{T})$ and $(Y, \mathscr{U})$ are spaces and $\mathscr{S}$ is a subbasis for the topology $\mathscr{U}$ of Y, then these three conditions on a mapping $f : X \to Y$ are equivalent:

(i) If U is open in Y then $f^{-1}[U]$ is open in X.
(ii) If $S \in \mathscr{S}$ then $f^{-1}[S]$ is open in X.
(iii) If C is closed in Y then $f^{-1}[C]$ is closed in X.

Proof Exercise. □

Definition

A mapping $f: X \to Y$ is said to be *continuous* if it satisfies the three equivalent conditions of lemma 9.6.1.

Definition

A bijective mapping $f: X \to Y$ is called a *homeomorphism* if both $f: X \to Y$ and $f^{-1}: Y \to X$ are continuous.

Remark 1 If $f: X \to Y$ and $g: Y \to Z$ are continuous mappings, then it is clear that $g \circ f: X \to Z$ is also continuous. Moreover, $\mathrm{id}_X: X \to X$ is trivially continuous.

Remark 2 A continuous bijective mapping need not be a homeomorphism (see exercise 3(c) below).

Lemma 9.6.2

If $(X, \mathscr{T})$ is a topological space and $Y \subseteq X$ then there is exactly one topology $\mathscr{T}_Y$ on Y such that if $f: Z \to X$ is a continuous mapping and $\mathrm{im}[f] \subseteq Y$ then the restriction $f: Z \to Y$ which makes the diagram

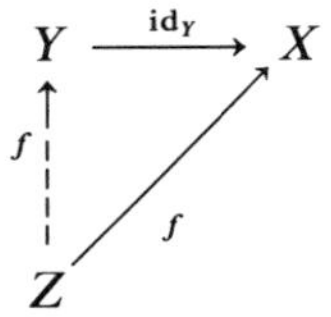

commute is continuous with respect to $\mathscr{T}_Y$.

Proof It is easy to check that $\mathscr{T}_Y = \{Y \cap U : U \in \mathscr{T}\}$ is the topology we require. □

Definition

The topology $\mathscr{T}_Y$ referred to in lemma 9.6.2 is called the *subspace* topology on Y and $(Y, \mathscr{T}_Y)$ is said to be a *subspace* of $(X, \mathscr{T})$.

We say that $f: X \to Y$ is an *embedding* if $f: X \to f[X]$ is a homeomorphism with respect to the subspace topology on $f[X]$.

Exercises

1. Prove lemma 9.6.1.
2. (a) Prove that every compact subspace of a Hausdorff space is closed. [Be careful not to use the axiom of choice in your proof.]

(b) Show that every subspace of a Hausdorff space is Hausdorff.

3. Let Top(X) denote the collection of all topologies on the collection X.
   (a) Show that if $\mathscr{T}, \mathscr{T}' \in \text{Top}(X)$ then $\mathscr{T} \subseteq \mathscr{T}'$ iff $\text{id}_X : X \to X$ is a continuous mapping from $(X, \mathscr{T}')$ to $(X, \mathscr{T})$.
   (b) Show that Top(X) is a complete lattice with respect to inclusion: its largest element is the *discrete* topology on X which consists of all the subcollections of X; its smallest element is the *indiscrete* topology which consists only of $\varnothing$ and X itself. The discrete topology on X is always Hausdorff but is compact iff X is finite; the indiscrete topology is always compact but is Hausdorff iff X has at most one element.
   (c) Give an example of a continuous bijective mapping which is not a homeomorphism.
   (d) Prove that a compact Hausdorff topology on a set X is a maximal element of the set of compact topologies on X and a minimal element of the set of Hausdorff topologies on X.
   (e) Give an example to show that the union of two topologies on a collection X need not be a topology on X. [Let X have three elements.]
4. (a) Show that a subspace A of a space X is irreducible iff $\text{Cl}(A)$ is an irreducible element of $\mathfrak{A}(X)$.
   (b) Show that every atom of $\mathfrak{A}(X)$ is of the form $\text{Cl}(x)$ for some $x \in X$ and that every collection of the form $\text{Cl}(x)$ is irreducible.
   (c) Show that a space X is irreducible iff every open subcollection is connected in the subspace topology.
5. A space X is said to be *hereditarily compact* if every subspace of X is compact.
   (a) Show that these three assertions are equivalent:
      (i) X is hereditarily compact.
      (ii) The lattice $\mathfrak{A}(X)$ of closed subsets of X is partially well ordered.
      (iii) Every open subset of X is compact in the subspace topology.
   (b) Show that if X is hereditarily compact then it is a finite union of irreducible subspaces.
   (c) Deduce that a space is hereditarily compact and Hausdorff iff it is finite and discrete.
6. Suppose that $(A, \leq)$ is an ordered set.
   (a) Prove that every closed bounded subset of A is compact iff A is boundedly complete. [*Necessity* If B is a subset of A which has an upper bound, let $B_x = \{y \in B : x \leq y\}$ for each $x \in B$ and observe that $\{B_x : x \in B\}$ has the finite intersection property. *Sufficiency* If B is a closed bounded subset of A and $\mathscr{D}$ is a set of closed subsets of B with the finite intersection property, let C be the set of all greatest lower bounds of elements of $\mathscr{D}$ and let $c = \sup C$; show that $c \in \bigcap \mathscr{D}$.]
   (b) Prove that A is connected (in the order topology) iff it is boundedly complete and strictly dense in itself. [*Necessity* Let B be a non-empty subset of A which has an upper bounded and let C be the set of all upper bounds for B in A; if C has no least element, show that C is a non-empty proper closed and open subcollection of A. *Sufficiency* If C is a non-empty proper closed and open subcollection of A, $b \in A - C$, $c \in C$, and (for the sake of argument) $b < c$, consider the greatest lower bound of C.]

Our next task is to define a topology on the product of a family o topological spaces in such a way that it has the characteristic propert we expect product objects to have.

Lemma 9.6.3

Suppose that $(X_i)_{i \in I}$ is a family of topological spaces. Then there is precisel one topology $\mathscr{T}$ on the collection $\Pi_{i \in I} X_i$ such that if $f_i : Y \to X_i$ is continuous mapping for each $i \in I$ then the unique mapping $f : Y \to \Pi_{i \in I} X$ which makes the diagram

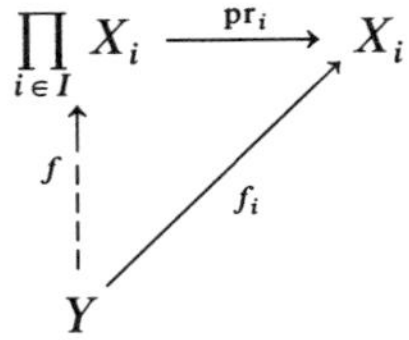

commute for all $i \in I$ is continuous.

Existence Let $\mathscr{T}$ be the topology on $\Pi_{i \in I} X_i$ generated by the collec tions $\mathrm{pr}_i^{-1}[U]$ where $i \in I$ and $U \in \mathscr{T}_i$. To check that the mapping $f : Y \to \Pi_{i \in I} X_i$ referred to in the proposition is continuous it is enough to sho that $f^{-1}[\mathrm{pr}_i^{-1}[U]]$ is open for each element $\mathrm{pr}_i^{-1}[U]$ of the subbasis fo $\mathscr{T}$. But

$$f^{-1}[\mathrm{pr}_i^{-1}[U]] = (\mathrm{pr}_i \circ f)^{-1}[U] = f_i^{-1}[U],$$

which is open since f_i is continuous.

Uniqueness If $\mathscr{T}'$ were another topology on $\Pi_{i \in I} X_i$ with the propert described in the proposition, then the identity function on $\Pi_{i \in I} X_i$ woul be the graph of a homeomorphism from $(\Pi_{i \in I} X_i, \mathscr{T})$ onto $(\Pi_{i \in I} X_i, \mathscr{T}'$ so that $\mathscr{T} = \mathscr{T}'$. [

Definition

The unique topology $\mathscr{T}$ referred to in lemma 9.6.3 is called the *produ topology* on $\Pi_{i \in I} X_i$ and the space $(\Pi_{i \in I} X_i, \mathscr{T})$ is called the *product* of th family $((X_i, \mathscr{T}_i))_{i \in I}$.

Remark 4 The collection $\{\mathrm{pr}^{-1}[U] : U \in \mathscr{T}_i$ for some $i \in I\}$ is a subbasis (sometim called the *canonical subbasis*) for the product topology. It generates a bas (the 'canonical' basis) which consists of the collections of the form $\Pi_{i \in I}$ where U_i is open in X_i for all $i \in I$ and $U_i = X_i$ for all but finitely many $i \in$

Remark 5 An alternative way to describe the product topology is to say that it the smallest topology on $\Pi_{i \in I} X_i$ with respect to which all the projectic mappings $\mathrm{pr}_j : \Pi_{i \in I} X_i \to X_j$ are continuous.

The product topology In 1923 Tietze considered the topology on $\Pi_{i \in I} X_i$ generated by *all* the collections $\Pi_{i \in I} U_i$ where each U_i is open in the corresponding X_i (the *box* topology). The potentially smaller topology which we have just defined was first mentioned by Tychonoff (1929) (in the particular case where each X_i is the unit interval). From an internal point of view the box topology is probably the more natural one to consider, but from an external point of view we are committed to the product topology as soon as we decide that the appropriate mappings between topological spaces are the continuous ones. (This observation can be made quite precise in the language of category theory.) However, the approach which category theory represents—that one should always ask not 'What are the objects?' but 'What are the morphisms?'—did not become widespread until the 1940s and was not apparent to Tychonoff, whose reason for defining the product topology was much more pragmatic: the box product of infinitely many compact spaces need not be compact, whereas we shall show in §9.7 that (assuming the axiom of choice) the Tychonoff product must be. Indeed it is for the converse pragmatic reason—it generates good counterexamples—that the box product has fairly recently (e.g. Rudin 1971) been the subject of renewed interest.

xercises

7. If $(X_i)_{i \in I}$ is a family of topological spaces, show that there exists exactly one topology on the set $\dot{\bigcup}_{i \in I} X_i$ with respect to which if Y is a space and if $f_i : X_i \to Y$ is a continuous mapping for each $i \in I$ then the unique mapping $f : \dot{\bigcup}_{i \in I} X_i \to Y$ which makes the diagram

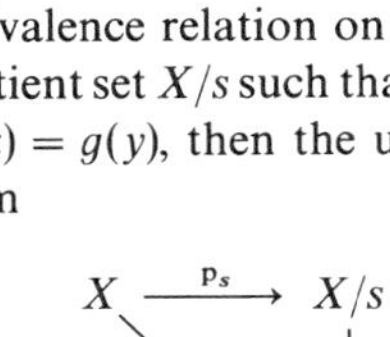

commute for all $i \in I$ is continuous.

8. (a) Show that if s is an equivalence relation on X, then there is exactly one topology $\mathscr{T}/s$ on the quotient set X/s such that if $g : X \to Y$ is a continuous mapping and $x\,s\,y \Rightarrow g(x) = g(y)$, then the unique mapping $g_s : X/s \to Y$ which makes the diagram

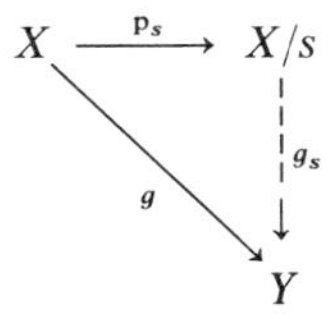

commute is continuous with respect to $\mathscr{T}/s$.

(b) If $g : X \to Y$ is a mapping, show that these two assertions are equivalent:

(i) There is an equivalence relation s on X such that $g_s : X/s \to Y$ is homeomorphism.
(ii) If $A \subseteq Y$ then $g^{-1}[A]$ is open in X iff A is open in Y.
(We say that $g : X \to Y$ is a *topological quotient mapping* if it satisfies the conditions.)

9. If $f : X \to Y$ is a continuous mapping, show that $U \mapsto f^{-1}[U]$ is a fra homomorphism $\mathfrak{O}(Y) \to \mathfrak{O}(X)$.
10. Suppose that $(X_i)_{i \in I}$ is a family of spaces and that $(a_i)_{i \in I} \in \Pi_{i \in I} X_i$.
    (a) If $Y_j = \{(x_i) \in \Pi_{i \in I} X_i : x_i = a_i \text{ for all } i \neq j\}$, show that the restricti $\mathrm{pr}_j | Y_j \to X_j$ of the jth projection mapping is a homeomorphism from t subspace Y_j to X_j.
    (b) If $Y = \{(x_i) \in \Pi_{i \in I} X_i : x_i = a_i \text{ for all but finitely many } i \in I\}$, show t $\mathrm{Cl}(Y) = \Pi_{i \in I} X_i$.
11. If $(X_i)_{i \in I}$ is a family of connected spaces, show that $\Pi_{i \in I} X_i$ is also connect [Use exercise 10.]
12. Suppose that $g : X \to Y$ is an open topological quotient mapping and th $\mathrm{Ker}[g]$ is a closed subcollection of $X \times X$. Show that Y is Hausdorff. mapping $g : X \to Y$ is *open* if $g[U]$ is open in Y whenever U is open in X.)

9.7 Ultrafilters in topology

We intend to show how a notion of convergence for filters of subcolle tions of a topological space may be defined. It will be convenient adapt our previous notation in order to exclude from consideration t improper filter of *all* subcollections of the space in question.

Definition

A proper filter in the lattice $\mathfrak{P}(A)$ will from now on be called simply *filter on A*; a maximal such filter will be called an *ultrafilter on A*. Similar filter bases in $\mathfrak{P}(A)$ which generate a filter on A are sometimes called *filt bases on A*. Arbitrary collections which generate a filter on A are call *filter subbases on A*.

The next two lemmas are simply translations into this notation of gener results about filters.

Lemma 9.7.1

(*a*) *A collection $\mathscr{S}$ of subcollections of A is a filter subbase on A iff it has t finite intersection property.*
(*b*) *A collection $\mathscr{B}$ of subcollections of A is a filter base on A iff*

$$B_1, B_2 \in \mathscr{B} \Rightarrow (\exists B \in \mathscr{B})(B \subseteq B_1 \cap B_2);$$

$$\varnothing \notin \mathscr{B}, \quad \mathscr{B} \neq \varnothing.$$

(*c*) *A collection $\mathscr{F}$ of subcollections of A is a filter on A iff*

$$F \in \mathscr{F}, \quad F \subseteq G \subseteq A \Rightarrow G \in \mathscr{F};$$

$$F_1, F_2 \in \mathscr{F} \Rightarrow F_1 \cap F_2 \in \mathscr{F};$$

$$\varnothing \notin \mathscr{F}, \quad A \in \mathscr{F}.$$

(*d*) *A collection* $\mathscr{U}$ *of subcollections of* A *is an ultrafilter on* A *iff it has the finite intersection property and*

$$B \subseteq A \Rightarrow (B \in \mathscr{U} \text{ or } A - B \in \mathscr{U}).$$

Proof Straightforward □

Lemma 9.7.2

If we assume the prime ideal property then every filter on a set A *is contained in at least one ultrafilter on* A.

Proof By the dual of part (ii) of theorem 9.3.5 every filter in $\mathfrak{P}(A)$ is contained in a prime filter. But $\mathfrak{P}(A)$ is Boolean and so by the dual of proposition 9.2.6 its prime filters are precisely the maximal proper filters, i.e. the ultrafilters on A. □

Lemma 9.7.3

Suppose that $f : A \to B$ *is a mapping. If* $\mathscr{A}$ *is a filter subbase [respectively a filter base, a filter, an ultrafilter] on* A *then* $\{f[A] : A \in \mathscr{A}\}$ *is a filter subbase [respectively a filter base, a filter, an ultrafilter] on* $f[A]$.

Proof Straightforward. □

Example 1 If X is a topological space and $x \in X$ then a subset C of X is called a *neighbourhood* of x if there exists an open set U in X such that $x \in U \subseteq C$. The set of all neighbourhoods of x is a filter on X called the *neighbourhood filter* of x and denoted $\mathscr{B}_x$.

Example 2 The cofinite subsets of ω (i.e. those with finite complements) form a filter on ω.

Example 3 If A is a non-empty subset of X then $\{B \subseteq X : A \subseteq B\}$ is a principal filter on X: it is an ultrafilter iff A has exactly one element.

Example 4 If (x_n) is a sequence in X then $\{\{x_n : n \geq n_0\} : n_0 \in \omega\}$ generates a filter on X which is said to be *determined by* the sequence (x_n).

Example 5 We can generalize example 4 as follows. A family indexed by a directed collection is called a *net*. The significance of this is that $(I, \leq)$ is directed iff $\{\uparrow i : i \in I\}$ is a filter base on I. So if $(x_i)_{i \in I}$ is a net in X then $\{\{x_j : j \geq i\} : i \in I\}$ is a filter base on X [lemma 9.7.3] and therefore generates a filter, which is said to be *determined by* the net $(x_i)_{i \in I}$.

Definition

A filter $\mathcal{F}$ on a topological space X is said to *converge* to the point $x \in X$ if $\mathcal{B}_x \subseteq \mathcal{F}$.

Exercises

1. A net $(x_i)_{i \in I}$ in a topological space X is said to *converge* to a point $x \in X$ if for every open $U \ni x$ there exists $i_0 \in I$ such that $x_i \in U$ for all $i \geq i_0$. (Thi evidently generalizes in a natural way the familiar definition of convergenc for sequences.) Show that this is the case iff the filter determined by $(x_i$ converges to x.
2. Assume the prime ideal property and let $\mathcal{F}$ be a filter on the topological spac X.
   (a) Show that $\mathcal{F}$ is the intersection of the ultrafilters containing it. [Us exercise 2(a) of §9.1.]
   (b) Show that $\mathcal{F}$ converges to a point x iff every ultrafilter containing $\mathcal{F}$ converges to x.

The notion of convergence The topological properties of subsets of a metric space may be characterized by properties of convergence o sequences. But the fact that sequences are countable makes their behaviou an inadequate guide to topological structure in more general settings. Thi is why Fréchet's attempt to define a general notion of 'topology' b means of the convergence of sequences was not eventually influential. Bu even after a notion of 'topology' with the right degree of generality ha been provided by Hausdorff (1914) it was some time before appropriat generalizations of convergence were defined—filters by Cartan (1937*a*, *b*) nets by Birkhoff (1937). Both these notions had appeared in mor restricted settings before: filters in $\mathbf{R}^n$ were mentioned by Caratheodor (1913); convergence of nets in $\mathbf{R}$ was studied by Moore and Smith (1922) The two approaches are essentially equivalent and for many years both were popular, filters following their popularization by Bourbaki (1940) nets particularly in America and among functional analysts.

Theorem 9.7.4

If X is a Hausdorff space then every ultrafilter on X converges to at mos one point of X; if X is small and we assume the prime ideal property the the converse is also true.

Necessity Suppose on the contrary that $\mathcal{U}$ is an ultrafilter on X whic converges to two distinct points x and y of X. Then there exist disjoin open collections $U \ni x$ and $V \ni y$ in X. Now $U \in \mathcal{B}_x \subseteq \mathcal{U}$ and $V \in \mathcal{B}_y \subseteq \mathcal{U}$ so that $\varnothing = U \cap V \in \mathcal{U}$. Contradiction.

Sufficiency Suppose that X is not Hausdorff. Then there exist distinc points x and y in X such that any open sets $U \ni x$ and $V \ni y$ must intersec It is easy to check that the subsets of X of the form $U \cap V$ for som open $U \ni x$ and $V \ni y$ form a set with the finite intersection property: it generates a filter on X [lemma 9.7.1(a)], which evidently converges to both x and y. Moreover, if we assume the prime ideal property then this filter is contained in an ultrafilter [lemma 9.7.2], which also converges to both x and y. Contradiction. □

heorem 9.7.5 (Cartan 1937*b*)

If X is a compact space then every ultrafilter on X converges to at least one point of X; if X is small and we assume the prime ideal property then the converse is also true.

Necessity Let $\mathscr{U}$ be an ultrafilter on X. If $A_0, \ldots, A_{n-1} \in \mathscr{U}$ then the intersection $\bigcap_{r \in n} \mathrm{Cl}(A_r)$ belongs to $\mathscr{U}$ and is therefore non-empty. In other words, $\{\mathrm{Cl}(A) : A \in \mathscr{U}\}$ has the finite intersection property and consequently (since X is compact) it has a common point x. Therefore every element of $\mathscr{U}$ intersects every element of $\mathscr{B}_x$. So there exists a filter on X containing both $\mathscr{U}$ and $\mathscr{B}_x$ [lemma 9.7.1(a)]. But this filter must be $\mathscr{U}$ itself because $\mathscr{U}$ is an ultrafilter. Thus $\mathscr{U}$ contains $\mathscr{B}_x$, i.e. $\mathscr{U}$ converges to x.

Sufficiency Let $\mathscr{A}$ be a set of closed subsets of X which has the finite intersection property. So $\mathscr{A}$ generates a filter on X [lemma 9.7.1(a)], which in turn is contained in an ultrafilter $\mathscr{U}$ on X [lemma 9.7.2], which converges to a point $x \in X$ by hypothesis. Now if $A \in \mathscr{A}$ then $A \in \mathscr{U}$, so that $X - A \notin \mathscr{U}$ and therefore $X - A \notin \mathscr{B}_x$; hence $x \notin X - A$ since $X - A$ is open, i.e. $x \in A$. In other words, $x \in \bigcap \mathscr{A}$. □

xercise

3. (a) If $\prod_{i \in I} X_i$ is Hausdorff and non-empty, show that each X_i is Hausdorff and non-empty. [Use exercise 10(a) of §9.6.]
   (b) If $\prod_{i \in I} X_i$ is compact and non-empty, show that each X_i is compact and non-empty.
   (c) If $\prod_{i \in I} X_i$ is connected and non-empty, show that each X_i is connected and non-empty.

heorem 9.7.6

If $(X_i)_{i \in I}$ is a family of Hausdorff spaces then $\prod_{i \in I} X_i$ is also Hausdorff.

roof Suppose that $x, y \in \prod_{i \in I} X_i$ and $x \neq y$. Then there exists an index $j \in I$ such that $\mathrm{pr}_j(x) \neq \mathrm{pr}_j(y)$. So since X_j is Hausdorff by hypothesis, there exist disjoint sets U and V open in X_j such that $\mathrm{pr}_j(x) \in U$ and $\mathrm{pr}_j(y) \in V$. Therefore $\mathrm{pr}_j^{-1}[U]$ and $\mathrm{pr}_j^{-1}[V]$ are disjoint open subsets of $\prod_{i \in I} X_i$ such that $x \in \mathrm{pr}_j^{-1}[U]$ and $y \in \mathrm{pr}_j^{-1}[V]$. Thus $\prod_{i \in I} X_i$ is Hausdorff. □

In contrast, the corresponding result for compact spaces depends crucially for its validity on the axiom of choice.

Lemma 9.7.7

If $\mathscr{F}$ is a filter on $\Pi_{i \in I} X_i$ such that $\{\mathrm{pr}_i[F] : F \in \mathscr{F}\}$ converges in X_i to for all $i \in I$ then $\mathscr{F}$ converges in $\Pi_{i \in I} X_i$ to $(x_i)_{i \in I}$.

Proof Suppose that $(x_i)_{i \in I} \in \mathrm{pr}_j^{-1}[U]$ with U open in X_j. It will be enough t show that $\mathrm{pr}_j^{-1}[U] \in \mathscr{F}$ since the collections of this form constitute subbase for the neighbourhood filter at (x_i). Now $x_j \in U$ and so

$$U \in \mathscr{B}_{x_j} \subseteq \{\mathrm{pr}_j[F] : F \in \mathscr{F}\}.$$

Hence there exists $F \in \mathscr{F}$ such that $U = \mathrm{pr}_j[F]$, so that

$$F \subseteq \mathrm{pr}_j^{-1}[\mathrm{pr}_j[F]] = \mathrm{pr}_j^{-1}[U]$$

and hence $\mathrm{pr}_j^{-1}[U] \in \mathscr{F}$ as required. [

Theorem 9.7.8

These three assertions are equivalent:

(*i*) *The axiom of choice.*

(*ii*) *If $(X_i)_{i \in I}$ is a small family of compact spaces then $\Pi_{i \in I} X_i$ is compact the product topology* (Tychonoff 1935).

(*iii*) *If I is a set and X is a small compact space then ${}^I X$ is compact in th product topology.*

(i) $\Rightarrow$ (ii) Suppose that $\mathscr{U}$ is an ultrafilter on $\Pi_{i \in I} X_i$. Then for each $i \in I$ th set $\{\mathrm{pr}_i[U] : U \in \mathscr{U}\}$ is an ultrafilter on X_i [lemma 9.7.3] and therefo converges [Cartan's theorem]. So by the axiom of choice there exists family $(x_i)_{i \in I}$ such that $\{\mathrm{pr}_i[U] : U \in \mathscr{U}\}$ converges to x_i for every $i \in$ and therefore $\mathscr{U}$ converges to $(x_i)_{i \in I}$ [lemma 9.7.7]. Hence $\Pi_{i \in I} X_i$ compact [Cartan's theorem].

(ii) $\Rightarrow$ (iii) Trivial.

(iii) $\Rightarrow$ (i) Suppose that $(A_i)_{i \in I}$ is a family of disjoint non-empty sets. Le us give the set $X = \bigcup_{i \in I} A_i$ the smallest topology such that A_i is close for every $i \in I$. It is straightforward to verify that with this topology X compact. So by hypothesis ${}^I X$ is compact. Now for each $i \in I$ the se $C_i = \{f \in {}^I X : f(i) \in A_i\}$ is evidently closed in ${}^I X$. Moreover, $\{C_i : i \in I$ has the finite intersection property [principle of finite choice]. Hence ther exists an element $f \in \bigcap_{i \in I} C_i$, which is obviously the graph of a mappin $f : I \to \bigcup_{i \in I} A_i$ such that $f(i) \in A_i$ for all $i \in I$. This proves the multiplica tive axiom, from which the axiom of choice follows [theorem 7.3.2]. [

If we combine the two results we have just proved, we can see tha assuming the axiom of choice, every product of compact Hausdorff space is also compact and Hausdorff. However, it is worth observing that thi claim is equivalent to the weaker prime ideal property.

heorem 9.7.9

These four assertions are equivalent:

(i) The prime ideal property.

(ii) The Boolean prime ideal property.

(iii) If $(X_i)_{i \in I}$ is a small family of compact Hausdorff spaces then $\Pi_{i \in I} X_i$ is compact and Hausdorff in the product topology.

(iv) ${}^I\mathbf{2}$ is compact in the product topology for every set I.

(i) $\Rightarrow$ (ii) Trivial.
(ii) $\Rightarrow$ (iii) The proof that $\Pi_{i \in I} X_i$ is compact goes as before except that the point $x_i \in X_i$ to which $\{\mathrm{pr}_i[U] : U \in \mathscr{U}\}$ converges is now unique [theorem 9.7.4] and we therefore do not need the axiom of choice to pick it. (We still use the prime ideal property for Boolean lattices *via* lemma 9.7.2.)
(iii) $\Rightarrow$ (iv) Trivial.
(iv) $\Rightarrow$ (i) Suppose that L is non-trivial. We intend to show that spec(L) is non-empty, i.e. that L has a prime ideal. For each finite set $A \subseteq L$ let

$$C_A = \{f \in {}^L\mathbf{2} : f|[A] \in \mathrm{spec}([A])\}$$

It is easy to verify that each such set C_A is closed in ${}^L\mathbf{2}$. Moreover, if $A_0, \ldots, A_{n-1}$ are finite subsets of L then $[A_0 \cup \cdots \cup A_{n-1}]$ is finite and non-trivial and therefore has a character f (say); if we extend f to the whole of L (by making it take the value 0 outside $[A_0 \cup \cdots \cup A_{n-1}]$, for example) then we obtain an element of $C_{A_0} \cap \cdots \cap C_{A_{n-1}}$. In other words, we have shown that $\{C_A : A \in \mathfrak{F}(L)\}$ has the finite intersection property. Since we are supposing that ${}^L\mathbf{2}$ is compact,

$$\varnothing \neq \bigcap_{A \in \mathfrak{F}(L)} C_A = \mathrm{spec}(L). \qquad \square$$

xercises

4. (a) Assuming the prime ideal property show that if $(X_i)_{i \in I}$ is a small family of compact Hausdorff spaces then $\Pi_{i \in I} X_i$ is compact in the box topology iff all but finitely many of the X_i have no more than one element.
   (b) A space X is said to be T_1 if $\{x\}$ is closed for all $x \in X$. Assuming the axiom of choice, extend (a) to small families of compact T_1 spaces.
5. Prove that these three assertions are equivalent:
   (i) The prime ideal property.
   (ii) If X is a small space such that every ultrafilter on X converges, then X is compact.
   (iii) If X is a small space and $\mathscr{S}$ is a subbasis for the topology of X such that every subset of $\mathscr{S}$ which covers X has a finite subset which covers X, then X is compact (Alexander 1939).

   [*(i)* $\Rightarrow$ *(ii)* Cartan's theorem. *(ii)* $\Rightarrow$ *(i)* Examine the proof of theorem 9.7.9. *(ii)* $\Rightarrow$ *(iii)* If $\mathscr{U}$ is an ultrafilter on X which does not converge, show that $\mathscr{S} - \mathscr{U}$ covers X. *(iii)* $\Rightarrow$ *(i)* If I is a set, show that ${}^I\mathbf{2}$ is compact by noting that $\{\mathrm{pr}_i^{-1}[0], \mathrm{pr}_i^{-1}[1] : i \in I\}$ is a subbasis for the product topology and that any subset of it which covers ${}^I\mathbf{2}$ has a finite subset which covers it. Then use theore 9.7.9]

6. Assume the axiom of countable choice.
   (a) Show that the neighbourhood filter of each point of ω_1 has a countab generating set but that the topology of ω_1 does not have a countable base.
   (b) Show that every increasing sequence in ω_1 converges.
   (c) Deduce that every sequence in ω_1 has a convergent subsequence.
   (d) Show that ω_1 is not compact.
7. Prove Szpilrajn's theorem assuming the prime ideal property. [If $(A, \leq)$ is partially ordered set, for each finite $F \subseteq A$ let V_F be the set of relations r on which induce on F a total ordering extending $\leq_F$. By using the compactne of ${}^{A\times A}\mathbf{2}$ with the product topology show that $\bigcap_{F \in \mathfrak{F}(A)} V_F \neq \varnothing$.]
8. Prove the colouring theorem assuming the prime ideal property.

9.8 Stone/Čech spaces

Lemma 9.8.1

If X is Hausdorff and A is a dense subcollection of X then

$$\operatorname{card}(X) \leq 2^{2^{\operatorname{card}(A)}}.$$

Proof It is clear that the mapping $X \to \mathfrak{P}(\mathfrak{P}(A))$ given by $x \mapsto \{U \cap A : U \in \mathscr{B}_x$ is injective. ☐

Lemma 9.8.2

If Y is Hausdorff and A is a dense subcollection of X then any tw continuous mappings $X \to Y$ which agree on A agree on X.

Proof Suppose that $f, g : X \to Y$ are continuous mappings which do not agre on X. So there exists $x \in X$ such that $f(x) \neq g(x)$. Since Y is Hausdor there exist disjoint open collections $U \ni f(x)$ and $V \ni g(x)$. Now x $f^{-1}[U] \cap g^{-1}[V]$ and so there exists $a \in f^{-1}[U] \cap g^{-1}[V] \cap A$. The $f(a) \in U$ and $g(a) \in V$. Consequently $f(a) \neq g(a)$ since U and V are disjoin Thus f and g do not agree on A. ☐

Theorem 9.8.3

These two assertions are equivalent:

(i) *The prime ideal property.*

(ii) *For every topological space X there exist a definite compact Hausdorf space βX and a continuous mapping $\mathrm{h} : X \to \beta X$ such that for ever compact Hausdorff space K and every continuous mapping $f : X \to K$ there exists a unique continuous mapping $\bar{f} : \beta X \to K$ which makes th diagram*

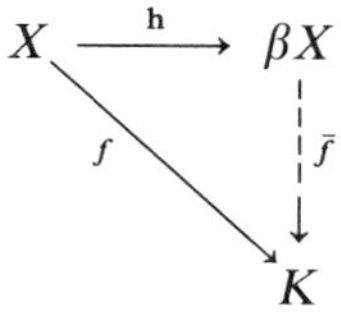

commute (Stone 1937*b*; Čech 1937).

(i) ⇒ (ii) Let $((K_\alpha, \mathscr{T}_\alpha, f_\alpha))_{\alpha \in A}$ be a family consisting of all the ordered triples $(K_\alpha, \mathscr{T}_\alpha, f_\alpha)$ such that $\mathscr{T}_\alpha$ is a compact Hausdorff topology on the subset K_α of $\mathfrak{P}(\mathfrak{P}(\mathfrak{P}(X)))$ and f_α is the graph of a continuous mapping $X \to K_\alpha$. Now define $\mathrm{h} : X \to \Pi_{\alpha \in A} K_\alpha$ by $\mathrm{h}(x) = (f_\alpha(x))_{\alpha \in A}$ and let βX be the closure of $\mathrm{h}[X]$ in $\Pi_{\alpha \in A} K_\alpha$. Then $\Pi_{\alpha \in A} K_\alpha$ is compact and Hausdorff [theorems 9.7.6 and 9.7.9] and so βX is compact and Hausdorff too.

Suppose now that K is a compact Hausdorff space and $f : X \to K$ is continuous. Then

$$\mathrm{card}(f[X]) < 2^{\mathrm{card}(X)}$$

and therefore

$$\begin{aligned}\mathrm{card}(\mathrm{Cl}(f[X])) &\leq 2^{2^{2^{\mathrm{card}(X)}}} \quad [\text{lemma 9.8.1}]\\ &= \mathrm{card}(\mathfrak{P}(\mathfrak{P}(\mathfrak{P}(X)))).\end{aligned}$$

So for some $\alpha_0 \in A$ there is a homeomorphism $g : K_{\alpha_0} \to \mathrm{Cl}(f[X])$ such that $g \circ f_{\alpha_0} = f$. If we let $\bar{f} = g \circ \mathrm{pr}_{\alpha_0} | \beta X \to K$ then

$$\bar{f} \circ \mathrm{h} = g \circ \mathrm{pr}_{\alpha_0} \circ \mathrm{h} = g \circ f_{\alpha_0} = f,$$

which is what we wanted. The uniqueness of $\bar{f}$ follows at once from lemma 9.8.2 since $\mathrm{h}[X]$ is by definition dense in βX.

(ii) ⇒ (i) Suppose that $(X_i)_{i \in I}$ is a small family of compact Hausdorff spaces. It will be sufficient [theorem 9.7.9] if we can show that the product space $X = \Pi_{i \in I} X_i$ is compact and Hausdorff. Now by assumption there exists for each $j \in I$ a unique continuous mapping $\overline{\mathrm{pr}}_j : \beta X \to X$ such that the diagram

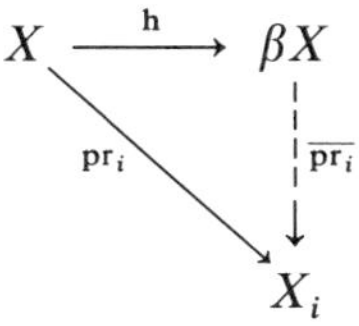

commutes. Define $g : \beta X \to X$ by $g(x) = (\overline{\mathrm{pr}}_i(x))_{i \in I}$. Then for any family $(x_j)_{j \in I} \in X$ we have

$$g(\mathrm{h}((x_j)_{j\in I})) = (\overline{\mathrm{pr}}_i(\mathrm{h}((x_j)_{j\in I})))_{i\in I}$$
$$= (\mathrm{pr}_i((x_j)_{j\in I}))_{i\in I}$$
$$= (x_i)_{i\in I}.$$

In other words $g \circ \mathrm{h} = \mathrm{id}_X$. But also both $\mathrm{h} \circ g$ and $\mathrm{id}_{\beta X}$ make the diagram

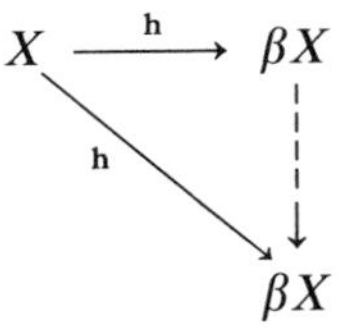

commute and so by the uniqueness property we must have $\mathrm{h} \circ g = \mathrm{id}_{\beta X}$. So the functions h and g are inverses of one another. Hence X is homeomorphic to βX and is therefore compact and Hausdorff. □

Remark 1 It is of course no accident that the proof of necessity in theorem 9.8.3 is similar in structure to the construction of Bool(I) in §7.5.

Definition

The compact Hausdorff space βX described in theorem 9.8.3 is called the *Stone/Čech space* of X (if it exists) and the mapping $\mathrm{h}: X \to \beta X$ is called the *Stone/Čech mapping*.

Proposition 9.8.4

If $\beta' X$ is another compact Hausdorff space and $h' : X \to \beta' X$ is a continuous mapping with the property described in theorem 9.8.3 *then there is exactly one homeomorphism $f: \beta X \to \beta' X$ such that $f \circ \mathrm{h} = \mathrm{h}'$.*

Proof Exercise. □

Definition

A topological space X is said to be *Tychonoff* if it can be embedded in a compact Hausdorff space.

Lemma 9.8.5

If X is a Tychonoff space then it is Hausdorff.

Proof Trivial. □

Remark 2 Assuming countable dependent choice, a small space X is Tychonoff iff it is Hausdorff and satisfies the following condition:

If $A \subseteq X$ is closed and $a \in X - A$, then there is a continuous mapping $f: X \to \overline{\mathbf{R}}$ such that $f(a) = -\infty$ and $f(x) = +\infty$ for all $x \in A$.

roposition 9.8.6

Assuming the prime ideal property, X is a Tychonoff space iff the Stone/Čech mapping $\mathrm{h}: X \to \beta X$ *is an embedding.*

Sufficiency Trivial.

Necessity Suppose that X is Tychonoff. Then there exists an embedding $f: X \to K$ of X into a compact Hausdorff space K. Now

$$\begin{aligned} \mathrm{h}(x) = \mathrm{h}(y) &\Rightarrow \bar{f}(\mathrm{h}(x)) = \bar{f}(\mathrm{h}(y)) \\ &\Rightarrow f(x) = f(y) \\ &\Rightarrow x = y \text{ since } f \text{ is injective,} \end{aligned}$$

i.e. $\mathrm{h}: X \to \beta X$ is injective. Moreover, if C is a closed subset of X then $f[C]$ is a closed subset of $f[X]$, so that $f[C] = D \cap f[X]$ for some closed subset D of K. Now it is straightforward to verify that $\mathrm{h}[C] = \bar{f}^{-1}[D] \cap \mathrm{h}[X]$, which is closed in $\mathrm{h}[X]$ since $\bar{f}$ is continuous. This shows that $\mathrm{h}: X \to \mathrm{h}[X]$ is a homeomorphism. □

emark 3 In the case when X is a Tychonoff space, it is usual to call the pair $(\beta X, \mathrm{h})$ (or just βX) the Stone/Čech *compactification* of X and to identity X with its image under h in βX. (In general, a compactification of X is a pair (K, i) such that K is compact, $i: X \to K$ is an embedding, and $i[X]$ is dense in K. It is not hard to show—see exercise 4 below—that every space has a compactification, but a space has a Hausdorff compactification iff it is Tychonoff.)

xercises

1. Prove proposition 9.8.4
2. Assume the prime ideal property.
   (a) If $p \in \beta\omega - \omega$, show that $\{A \cap \omega : A \in \mathscr{B}_p\}$ is a non-principal ultrafilter on ω and $\mathscr{B}_p$ does not have a countable base. [Use exercise 2(c) of §9.2.]
   (b) Show that every compact Hausdorff separable space is a continuous image of $\beta\omega$.
3. Prove that if X is a topological space then there exist a Hausdorff space $\mathrm{T}_2(X)$ and a continuous mapping $\mathrm{t}_2: X \to \mathrm{T}_2(X)$ such that for any Hausdorff space Y and continuous mapping $f: X \to Y$ there is exactly one continuous mapping $\bar{f}: \mathrm{T}_2(X) \to Y$ making the diagram

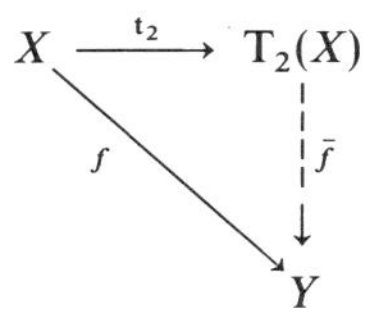

commute.

4. Suppose that $(X, \mathscr{T})$ is a topological space. If X is compact, let $X^+ = X$; if no let $X^+ = X \cup \{\infty\}$, where ∞ is a definite object not in X (such as X itself) an let

$$\mathscr{T}^+ = \mathscr{T} \cup \{X^+ - C : C \text{ is a compact closed subcollection of } X\}.$$

(a) Show that (X^+, id_X) is a compactification of X (called the *Alexandro one-point compactification*).
(b) Show that X^+ is Hausdorff iff X is locally compact and Hausdorff.
(c) Deduce that every locally compact Hausdorff space is Tychonoff.

9.9 The topology of the spectrum

In §9.3 we saw that (assuming the prime ideal property) every sma distributive lattice L can be represented as a sublattice of the power se $\mathfrak{P}(\mathrm{spec}(L))$. We are now in a position to define a structure on $\mathrm{spec}(L$ which determines which sublattice it is: the structure in question is th topology on $\mathrm{spec}(L)$ generated by $\{\mathrm{V}(a) : a \in L\}$.

Definition

Define a mapping $\mathscr{V} : \mathfrak{P}(L) \to \mathfrak{P}(\mathrm{spec}(L))$ by letting

$$\mathscr{V}(A) = \bigcup_{a \in A} \mathrm{V}(a) = \{f \in \mathrm{spec}(L) : (\exists a \in A)(f(a) = \top)\}.$$

Lemma 9.9.1

$\mathscr{V}[\mathfrak{P}(L)]$ is a topology on $\mathrm{spec}(L)$.

Proof Trivial. □

Definition

$\mathscr{V}[\mathfrak{P}(L)]$ is called the *spectral* topology.

Whenever we refer to $\mathrm{spec}(L)$ as a topological space we shall mean it t be endowed with this topology. Here are some other descriptions of it.

Lemma 9.9.2

(a) $\mathscr{V}[\mathrm{Idl}(L)]$ is the spectral topology on $\mathrm{spec}(L)$.
(b) $\mathrm{V}[L]$ is a basis for the spectral topology on $\mathrm{spec}(L)$.
(c) If A is a subcollection of L which generates it as a lattice, then $\mathrm{V}[A$ is a subbasis for the spectral topology on $\mathrm{spec}(L)$.

Proof Exercise. □

Remark 1 We can obtain another description of the spectral topology as follows Suppose that we let $\mathbf{2}_s$ denote the set $\{\bot, \top\}$ with the upper topology (se that $\{\top\}$ is open but $\{\bot\}$ is not). Then the product topology on ${}^{L}\mathbf{2}_s$ gives rise to a subspace topology on spec(L) (since spec(L), you will recall, consists of the characters $L \to \mathbf{2}$). It is easy to check that this subspace topology is just the spectral topology.

Exercises

1. Prove lemma 9.9.2.
2. (a) Show that $\mathscr{V}: \mathrm{Idl}(L) \to \mathfrak{O}(\mathrm{spec}(L))$ is a surjective frame homomorphism.
   (b) Assuming the prime ideal property, show that it is an isomorphism.

Example 1 The canonical bijection ${}^{A}\mathbf{2} \to \mathrm{spec}(\mathrm{Bool}(A))$ (example 4 of §9.2) maps the canonical subbasis $\{\mathrm{pr}_a^{-1}[\bot], \mathrm{pr}_a^{-1}[\top] : a \in A\}$ for the product topology on ${}^{A}\mathbf{2}$ ($\mathbf{2}$ having the discrete topology as usual) onto $\{\mathrm{V}(\mathbf{X}_a), \mathrm{V}(\mathbf{X}_a^*) : a \in A\}$, which is a subbasis for the spectral topology [lemma 9.9.2(c)]. It follows that the spectrum of Bool(A) is homeomorphic to ${}^{A}\mathbf{2}$.

Example 2 If we identify spec($\mathfrak{FC}(A)$) with A^+ by means of the bijection referred to in example 3 of §9.2, then each element of A is isolated in the spectral topology and the open neighbourhoods of ∞ are the cofinite subcollections of A^+. In other words, the spectrum of $\mathfrak{FC}(A)$ is homeomorphic to the Alexandroff one-point compactification of the topological space obtained by giving A the discrete topology.

If $f, g \in \mathrm{spec}(L)$ then $f \leq g$ iff $g \in \mathrm{Cl}(f)$. In other words, the spectral topology is a lower topology on the partially ordered set spec(L). It therefore encodes at least as much information about L as the partial ordering does: the next theorem shows that if we assume the prime ideal property then it actually encodes more, since it enables us to do what we failed to do with the partial ordering, namely retrieve the original lattice L.

Theorem 9.9.3

These five assertions are equivalent:

(i) *The prime ideal property.*

(ii) *If L is small, then the image of the canonical mapping* $\mathrm{V}: L \to \mathfrak{P}(\mathrm{spec}(L))$ *is the set* $\mathfrak{RO}(\mathrm{spec}(L))$ *consisting of the subsets of* spec(L) *which are both compact and open in the spectral topology* (Stone 1937*a*).

(iii) *If L is small, then* spec(L) *is compact in the spectral topology;*

(iv) *If L is small and Boolean, then* spec(L) *is compact in the spectral topology.*

(v) *If I is a set then* spec(Bool(I)) *is compact in the spectral topology.*

(i) $\Rightarrow$ (ii) For each $a \in L$ the set V(a) is certainly open in spec(L). To show that it is compact, suppose that $\{\mathrm{V}(x) : x \in B\}$ is a covering of V(a) by elements of the basis $\{\mathrm{V}(x) : x \in L\}$. Now if $a \notin {\downarrow}[B]$ then by the separation property there exists $f \in \mathrm{spec}(L)$ such that $f(a) = \top$ and $f[B] = \{\bot\}$; in particular, $f \in \mathrm{V}(a)$, but $f \notin \bigcup_{x \in B} \mathrm{V}(x) \supseteq \mathrm{V}(a)$, which is a contradiction. Thus $a \in \downarrow[B]$ and so there exist $x_0, \ldots, x_{n-1} \in B$ such th $a \leq x_0 \sqcup \cdots \sqcup x_{n-1}$. Hence

$$\mathrm{V}(a) \subseteq \mathrm{V}(x_0) \cup \cdots \cup \mathrm{V}(x_{n-1}).$$

This shows that $\mathrm{V}(a)$ is compact.

Conversely, suppose that U is a subset of spec(L) which is both com pact and open in the spectral topology. Since U is open, there exis $B \subseteq L$ such that $U = \bigcup_{x \in B} \mathrm{V}(x)$; since U is compact, B has a finite subs $\{x_0, \ldots x_{n-1}\}$ such that

$$U = \mathrm{V}(x_0) \cup \cdots \cup \mathrm{V}(x_{n-1}) = \mathrm{V}(x_0 \sqcup \cdots \sqcup x_{n-1}).$$

This shows that U is in the image of V.

(ii) ⇒ (iii) ⇒ (iv) ⇒ (v) Trivial.

(v) ⇒ (i) If I is a set, then ${}^I\mathbf{2}$ is homeomorphic to spec(Bool(I)) (se example 1) and is therefore compact by hypothesis. The prime ide property follows [theorem 9.7.9]. □

Let us emphasize in particular that if we assume the prime ideal propert then a small distributive lattice is isomorphic to the set of compact ope subsets of its spectrum. So the topology of the spectrum has encoded i it enough information to enable us to retrieve (an isomorphic copy of the original lattice. It is therefore to be expected that we should be abl to characterize properties of the lattice by topological properties of th spectrum. For instance:

Proposition 9.9.4

(a) *If L is Boolean then* spec(L) *is Hausdorff.*

(b) *If we assume the prime ideal property, then the converse is also tru (provided that L is small).*

Proof (a) Suppose that $f, g \in$ spec(L) and $f \neq g$. So there exists $a \in L$ such tha $f(a) \neq g(a)$. Let us suppose for the sake of argument that $f(a) = \top$ and $g(a) = \bot$, so that $g(a^*) = g(a)^* = \top$. Thus $f \in \mathrm{V}(a)$ and $g \in \mathrm{V}(a^*)$ Moreover,

$$\mathrm{V}(a) \cap \mathrm{V}(a^*) = \mathrm{V}(a \sqcap a^*) = \mathrm{V}(\bot) = \varnothing.$$

This shows that spec(L) is Hausdorff.

(b) If we assume the prime ideal property then L is isomorphic to th lattice $\mathfrak{KO}$(spec(L)) [theorem 9.9.3], and if spec(L) is Hausdorff the $\mathfrak{KO}$(spec(L)) is evidently Boolean since it is identical to the se $\mathfrak{BO}$(spec(L)) of all the subsets of spec(L) which are simultaneousl closed and open. □

Proposition 9.9.5

(a) *If L is countable, then* spec(L) *is compact and has a countable basis.*
(b) *If we assume the prime ideal property then the converse is also true (provided that L is small).*

Proof (a) If L is countable then the proof that spec(L) is compact no longer requires us to assume the prime ideal property since for countable lattices this is provable in basic set theory [proposition 9.3.3]. Further $\{V(a) : a \in L\}$ is then by definition a countable basis for the spectral topology.

(b) Now suppose that spec(L) is compact and has a countable basis $\{U_n : n \in \omega\}$. If $K \in \mathfrak{KO}(\text{spec}(L))$ and $B(K) = \{n \in \omega : U_n \subseteq K\}$ then $K = \bigcup_{n \in B(K)} U_n$ since $\{U_n : n \in \omega\}$ is a basis for the spectral topology. Hence since K is compact there is a finite set $F(K) \subseteq B(K)$ such that $K = \bigcup_{n \in F(K)} U_n$: indeed we can specify $F(K)$ uniquely by requiring it to occur as early as possible in some well-ordering on $\mathfrak{F}(\omega)$ chosen in advance. We thus define an injective mapping $F : \mathfrak{KO}(\text{spec}(L)) \to \mathfrak{F}(\omega)$, from which it follows that

$$\begin{aligned} \text{card}(L) &= \text{card}(\mathfrak{KO}(\text{spec}(L))) \quad [\text{theorem 9.9.3}] \\ &\leq \text{card}(\mathfrak{F}(\omega)) \\ &= \aleph_0. \end{aligned}$$

□

Remark 2 The assumption of the prime ideal property in results of this kind is inevitable: without it the spectra of non-trivial distributive lattices may very well be empty and their topologies will then tell us nothing about the properties of the lattices they are derived from.

Exercises

3. Show, assuming the prime ideal property, that the spectrum of $\mathfrak{P}(A)$ is homeomorphic to the Stone/Cech compactification of the space obtained by giving A the discrete topology.
4. Suppose that L is a frame.
   (a) Show that the image $V^{\uparrow}[L]$ of the canonical mapping $V^{\uparrow} : L \to \mathfrak{P}(\text{pt}(L))$ is $\mathfrak{O}(\text{pt}(L))$, where pt($L$) has the subspace topology inherited from the spectral topology on spec(L).
   (b) Show that these three assertions are equivalent:
      (i) L is spatial;
      (ii) $V^{\uparrow} : L \to \mathfrak{O}(\text{pt}(L))$ is an isomorphism;
      (iii) there exists a space X such that L is isomorphic to $\mathfrak{O}(X)$.
   (c) Show that pt(L) is homeomorphic to pt($\mathfrak{O}$(pt(L))).
5. Suppose that X is a space.
   (a) Show that the mapping $s : X \to \text{pt}(\mathfrak{O}(X))$ given by $s(x)(U) = c_U(x)$ is continuous.

(b) Show that these three assertions are equivalent:
  (i) $s : X \to \mathrm{pt}(\mathfrak{O}(X))$ is a homeomorphism;
  (ii) there exists a frame L such that X is isomorphic to $\mathrm{pt}(L)$;
  (iii) for every closed irreducible subspace A of X there exists a unique $a \in X$ such that $\mathrm{Cl}(a) = A$. X is said to be *sober* in these circumstances.

(c) Show that $\mathrm{pt}(\mathfrak{O}(X))$ is sober and that if Y is sober and $f : X \to Y$ is a continuous mapping then there exists a unique continuous mapping $\bar{f} : \mathrm{pt}(\mathfrak{O}(X)) \to Y$ which makes the diagram

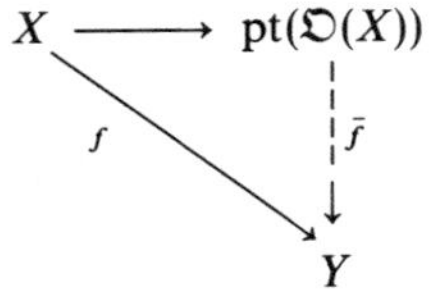

commute.

(d) Show that $\mathfrak{O}(X)$ is isomorphic to $\mathfrak{O}(\mathrm{pt}(\mathfrak{O}(X)))$.

9.10 Exercises on commutative rings

Throughout these exercises R denotes a commutative ring with identity element.

1. For each ideal $\mathfrak{a}$ in R we let $\sqrt{\mathfrak{a}}$ denote the *radical* of $\mathfrak{a}$, i.e. the set $\{x \in R : x^n \in$ for some $n > 0\}$.
   (a) Show that $\sqrt{\mathfrak{a}}$ is an ideal in R containing $\mathfrak{a}$ and that $\sqrt{\sqrt{\mathfrak{a}}} = \sqrt{\mathfrak{a}}$.
   (b) If $\mathfrak{a}^r \subseteq \mathfrak{b}$ for some $r > 0$ then $\sqrt{\mathfrak{a}} \subseteq \sqrt{\mathfrak{b}}$.
   (c) $\sqrt{(\mathfrak{a}\mathfrak{b})} = \sqrt{(\mathfrak{a} \cap \mathfrak{b})} = \sqrt{\mathfrak{a}} \cap \sqrt{\mathfrak{b}}$.
2. $\mathfrak{a}$ is called a *radical ideal* of R if $\sqrt{\mathfrak{a}} = \mathfrak{a}$.
   (a) Show that the set $\mathrm{Rid}(R)$ of radical ideals of R is a coherent frame with respect to inclusion.
   (b) Show that the compact elements of $\mathrm{Rid}(R)$ are precisely the finitely generated radical ideals of R.
   (c) Show that the coirreducible elements of $\mathrm{Rid}(R)$ are precisely the prime ideals of R and the coatoms are the maximal ideals.
   (d) R is said to be a *Boolean* ring if $x^2 = x$ for all $x \in L$. Show that every Boolean lattice L becomes a Boolean ring L_r when it is given the operations

$$xy = x \sqcap y$$

$$x + y = (x \sqcup y) \sqcap (x* \sqcup y*)$$

   (e) Show that these four assertions are equivalent:
     (i) the prime ideal property;
     (ii) every coherent frame is spatial;
     (iii) every non-trivial compact frame has a point;
     (iv) every commutative ring with identity has a prime ideal.

   [*(i)* ⇒ *(iii)* See exercise 5(b) of §9.3. *(i)* ⇒ *(ii)* Use exercise 2(b) of §9.9 and exercise 5(b) of §9.3. *(ii)* ⇒ *(iv)* Use (c). *(iii)* ⇒ *(iv)* Ditto. *(iv)* ⇒ *(i)* Use (d) and theorem 9.7.9.]

3. The distributive lattice $\mathfrak{R}(\mathrm{Rid}(R))$ is called the *reticulation* of the ring R and denoted $\mathrm{L}(R)$. There is a canonical mapping $\mathrm{D} : R \to \mathrm{L}(R)$ given by $\mathrm{D}(a) = \sqrt{(a)}$.
   (a) Show that

$$\mathrm{D}(0) = \bot; \quad \mathrm{D}(1) = \top;$$

$$\mathrm{D}(ab) = \mathrm{D}(a) \sqcap \mathrm{D}(b);$$

$$\mathrm{D}(a + b) \leq \mathrm{D}(a) \sqcup \mathrm{D}(b).$$

   (b) Suppose that $f : R \to L$ is a mapping such that

$$f(0) = \bot; \quad f(1) = \top;$$

$$f(ab) = f(a) \sqcap f(b);$$

$$f(a + b) \leq f(a) \sqcup f(b).$$

   Show that there is exactly one lattice homomorphism $\bar{f} : \mathrm{L}(R) \to L$ which makes the diagram

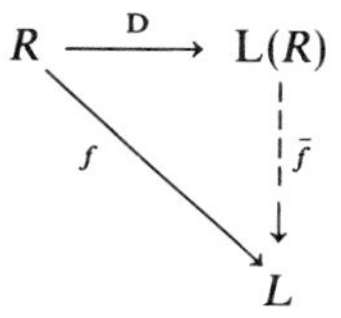

   commute.
   (c) Show that these three assertions are equivalent:
      (i) R is a Boolean ring;
      (ii) the canonical mapping $\mathrm{D} : R \to \mathrm{L}(R)$ is injective;
      (iii) the canonical mapping $\mathrm{D} : R \to \mathrm{L}(R)$ is bijective.
   (d) Show that if R is a Boolean ring then $\mathrm{L}(R)$ is a Boolean lattice and the associated Boolean ring $\mathrm{L}(R)_r$ is isomorphic to R.

References

Aczel, P. (1988). *Non-well-founded sets*, CSLI Lecture Notes, **14**. Center for the Study of Language and Information, Leland Stanford Junior University.

Alexander, J. W. (1939). Ordered sets, complexes, and the problem of bicompactification. *Proc. Natl. Acad. Sci. USA*, **25**, 296–8.

Baire, R. *et al.* (1905). Cinq lettres sur la théorie des ensembles. *Bull. Soc. Math. France*, **33**, 261–73. English translation in Moore (1982, pp. 311–20).

Banach, S. and Tarski, A. (1924). Sur la décomposition des ensembles de points en parties respectivement congruentes. *Fundam. Math.*, **6**, 262.

Banaschewski, B. (1985). Prime elements from prime ideals. *Order*, **2**, 211–13.

Barker, S. F. (1969). Realism as a philosophy of mathematics. In Bulloff *et al.* (1969, pp. 1–9).

Benacerraf, P. and Putnam, H. (eds) (1983). *Philosophy of mathematics. Selected readings*, (2nd edn). Cambridge University Press.

Bergmann, G. (1929). Zur Axiomatik der Elementargeometrie. *Monatsh. Math. Phys.*, **36**, 269–84.

Bernays, P. (1942). A system of set theory III. *J. Symb. Logic*, **7**, 65–89.

Bernoulli, J. (1686). Demonstratio rationum.... *Acta eruditorum*, 360–1. Reprinted in J. Bernoulli, *Opera*, Vol. 1, Geneva, 1744, pp. 282–3.

Bernstein, F. (1901). Untersuchungen aus der Mengenlehre. Doctoral dissertation. University of Göttingen. Reprinted with alterations as Bernstein (1905*a*).

Bernstein, F. (1905*a*). Untersuchungen aus der Mengenlehre. *Math. Ann.*, **61**, 117–55.

Bernstein, F. (1905*b*). Über die Reihe der transfiniten Ordnungszahlen. *Math. Ann.*, **60**, 187–93.

Bettazzi, R. (1896). Gruppi finiti ed infiniti di enti. *Atti Accad. Sci. Torino Cl. Sci. Fis. Mat. Nat.*, **31**, 506–12.

Birkhoff, G. (1933). On the combination of subalgebras. *Proc. Cambridge Philos. Soc.*, **29**, 441–64.

Birkhoff, G. (1937). Moore-Smith convergence in general topology. *Ann. Math. (2)*, **38**, 39–56.

Blair, C. E. (1977). The Baire category theorem implies the principle of dependent choices. *Bull. Acad. Pol. Sci., Sér. Sci. Math., Astron. Phys.*, **25**, 933–4.

Bochner, S. (1928). Fortsetzung Riemannscher Flaschen. *Math. Ann.*, **98**, 406–21.

Bolzano, B. (1851). *Paradoxien des Unendlichen*, (ed. Fr. Přihonský). Reclam, Leipzig. English translation Bolzano (1950).

Bolzano, B. (1950). *Paradoxes of the infinite*, (trans. and ed. D. A. Steele). Routledge & Kegan Paul, London.

Borel, E. (1898). *Leçons sur la théorie des fonctions.* Gauthiers-Villars, Paris.

Borel, E. (1905). Quelques remarques sur les principes de la théorie des ensembles. *Math. Ann.*, **60**, 194–5.

Bourbaki, N. (1939). *Théorie des ensembles* (*Fascicule de résultats*), Actualités scientifiques et industrielles, **846**. Hermann, Paris.

Bourbaki, N. (1940). *Topologie générale*, Ch. I et II, Actualités scientifiques et industrielles, **858**. Hermann, Paris.

Bourbaki, N. (1949*a*). Sur le théorème de Zorn. *Arch. Math.*, **2**, 434–7.

Bourbaki, N. (1949*b*). Foundations of mathematics for the working mathematician. *J. Symb. Logic*, **14**, 1–8.

Bourbaki, N. (1954). *Théorie des ensembles*, Ch. I et II, Actualités scientifiques et industrielles, **1212**. Hermann, Paris.

Bulloff, J. J., Holyoke, T. C., and Hahn, S. W. (eds) (1969). *Foundations of mathematics: Symposium papers commemorating the sixtieth birthday of Kurt Gödel*. Springer, Berlin.

Burali-Forti, C. (1896). Sopra un teorema del sig. G. Cantor. *Atti Accad. Sci. Torino Cl. Sci. Fis. Mat. Nat.*, **32**, 229–37.

Burali-Forti, C. (1897*a*). Una questione sui numeri transfiniti. *Rend. Circ. Mat. Palermo*, **11**, 154–64. English translation in van Heijenoort (1967, pp. 104–11).

Burali-Forti, C. (1897*b*). Sulle classi ben ordinate. *Rend. Circ. Mat. Palermo*, **11**, 260. English translation in van Heijenoort (1967, pp. 111–12).

Burali-Forti, C. (1906). Letter to Couturat. In Couturat (1906).

Campbell, P. J. (1978). The origins of Zorn's lemma. *Hist. Math.*, **5**, 77–89.

Cantor, G. (1873*a*). Letter to Dedekind, 29 November 1873. In Noether and Cavaillès (1937, pp. 12–13).

Cantor, G. (1873*b*). Letter to Dedekind, 7 December 1873. In Noether and Cavaillès (1937, pp. 14–15).

Cantor, G. (1877). Letter to Dedekind, 29 June 1877. In Noether and Cavaillès (1937, p. 34).

Cantor, G. (1878). Ein Beitrag zur Mannigfaltigkeitslehre. *J. reine angew. Math.*, **84**, 242–58. Reprinted in Cantor (1932, pp. 119–33).

Cantor, G. (1882). Über unendliche, lineare Punktmannichfaltigkeiten III. *Math. Ann.*, **20**, 113–21. Reprinted in Cantor (1932, pp. 149–57).

Cantor, G. (1883*a*). Über unendliche, lineare Punktmannichfaltigkeiten V. *Math. Ann.*, **21**, 545–91. Reprinted as Cantor (1883*b*); also in Cantor (1932).

Cantor, G. (1883*b*) *Grundlagen einer allgemeinen Mannigfaltigkeitslehre. Ein mathematisch-philosophischer Versuch in der Lehre des Unendlichen*. Leipzig.

Cantor, G. (1887–8). Mitteilungen zur Lehre vom Transfiniten. *Z. Philos. philos. Krit.*, **91**, 81–125, **92**, 240–65. Reprinted in Cantor (1932, pp. 378–439).

Cantor, G. (1892). Über eine elementare Frage der Mannigfaltigkeitslehre. *Jahresber. Dtsch. Math.-Ver.*, **1**, 75–8. Reprinted in Cantor (1932, pp. 278–81).

Cantor, G. (1895, 1897*a*). Beitrage zur Bergrundung der transfiniten Mengenlehre. *Math. Ann.*, **46**, 481–512, **49**, 207–46. Reprinted in Cantor (1932, pp. 282–356); English translation in Cantor (1915).

Cantor, G. (1897*b*). Letter to Hilbert, 26 September 1897. In Purkert and Ilgauds (1987, pp. 224–6).

Cantor, G. (1897*c*). Letter to Hilbert, 2 October 1897. In Purkert and Ilgauds (1987, pp. 226–7).

Cantor, G. (1899*a*). Letter to Dedekind, 28 July 1899. Inaccurately published in Cantor (1932, pp. 443–7). English translation in van Heijenoort (1967, pp. 113–17); see Grattan-Guinness (1974).

Cantor, G. (1899*b*). Letter to Dedekind, 3 August 1899. Inaccurately published in Cantor (1932, pp. 443–7). English translation in van Heijenoort (1967, pp. 113–17); see Grattan-Guinness (1974).

Cantor, G. (1903). Letter to Jourdain, 4 November 1903. In Grattan-Guinness (1971, pp. 116–17).

Cantor, G. (1915). *Contributions to the founding of the theory of transfinite numbers*, (trans. and ed. P. E. B. Jourdain). Open Court, Chicago. Reprinted by Dover, New York, 1955.

Cantor, G. (1932). *Gesammelte Abhandlungen*, (ed. E. Zermelo). Springer, Berlin. Reprinted by Olms, Hildesheim, 1962.

Caratheodory, C. (1913). Über die Begrenzung einfach zusammenhängender Gebiete. *Math. Ann.*, **73**, 323–70.

Cartan, H. (1937*a*). Théorie des filtres. *C. R. Acad. Sci.*, **205**, 595–8. Reprinted in Cartan (1979, pp. 953–6).

Cartan H. (1937*b*). Filtres et ultrafiltres. *C. R. Acad. Sci.*, **205**, 777–9. Reprinted in Cartan (1979, pp. 957–9).

Cartan, H. (1979). *Oeuvres*, Vol. III, (ed. R. Remmert and J.-P. Serre). Springer, Berlin.

Cassinet, J. (1980). L'axiome multiplicatif et autres formes de l'axiome du choix chez Russell et Whitehead. *Arch. Int. Hist. Sci.*, **30**, 69–85.

Čech, E. (1937). On bicompact spaces. *Ann. Math.* (2), **38**, 823–44.

Cohen, P. J. (1963–4). The independence of the continuum hypothesis. *Proc. Natl. Acad. Sci. USA*, **50**, 1143–8, **51**, 105–10.

Cohen, P. J. (1966). *Set theory and the continuum hypothesis.* Benjamin, Reading, Mass.

Couturat, L. (1906). Pour la logistique. *Rev. Métaphys. Morale*, **14**, 208–50.

Davis, M. (ed.) (1965). *The undecidable.* Raven, New York.

Dedekind, R. (1888). *Was sind und was sollen die Zahlen?*. Vieweg, Braunschweig. English translation as Dedekind (1901).

Dedekind, R. (1897). Letter to Dedekind, 29 August 1897, and accompanying proof. In Grattan-Guinness (1974, pp. 129–30) and Cantor (1932, p. 449).

Dedekind, R. (1901). *Essays on the theory of numbers*, (trans. W. W. Beman). Open Court, New York. Reprinted by Dover, New York, 1963.

Dedekind, R. (1930–2). *Gesammelte mathematische Werke*, 3 Vols, (ed. R. Fricke, E. Noether, and O. Öre). Vieweg, Braunschweig.

Dilworth, R. P. (1945). Lattices with unique complements. *Trans. Am. Math. Soc.*, **57**, 123–54.

Dixon, A. C. (1905). On 'well-ordered' aggregates. *Proc. London Math. Soc.* (2), **4**, 18–20.

Dugac, P. (1976). *Richard Dedekind et les fondements des mathématiques.* Vrin, Paris.

Euclid of Alexandria (*c*.300 BC). *Elements.* English translation as Euclid (1908).

Euclid of Alexandria (1908). *The thirteen books of Euclid's Elements*, 3 Vols, (trans. and ed. T. L. Heath). Cambridge University Press. Reprinted by Dover, New York, 1956.

Feferman, S. (1965). Some applications of the notions of forcing and generic sets. *Fundam. Math.*, **56**, 325–45.

Feferman, S. and Levy, A. (1963). Independence results in set theory by Cohen's method. *Not. Am. Math. Soc.*, **10**, 593.

Fraenkel, A. A. (1922*a*). Zu den Grundlagen der Cantor-Zermeloschen Mengenlehre. *Math. Ann.*, **86**, 230–7.

Fraenkel, A. A. (1922*b*). Über den Begriff 'definit' und die Unabhängigkeit des Auswahlaxioms. *Sitzungsber. Preuss. Akad. Wiss., Phys.-math. Kl.*, 253–7. English translation in van Heijenoort (1967, pp. 284–9).

Fraenkel, A. A. (1967). *Lebenskreise: Aus den Erinnerung eines jüdischen Mathematikers.* Deutsche Verlags-Anstalt, Stuttgart.

Fraenkel, A. A., Bar-Hillel, Y., and Levy, A. (1973). *Foundations of set theory*, (2nd revised edn), Studies in logic and the foundations of mathematics, **67**. North-Holland, Amsterdam.

Frege, G. (1879). *Begriffschrift, eine der arithmetischen nachgebildete Formelspruche des reinen Denkens.* Nebert, Halle a.S. New edn, (ed. I. Angelelli), Olms, Hildesheim, 1964; English translation in van Heijenoort (1967, pp. 5–82); also in Frege (1972).

Frege, G. (1884). *Die Grundlagen der Arithmetik.* Breslau. English translation as Frege (1980*c*).

Frege, G. (1893, 1903). *Grundgesetze der Arithmetik, begriffschriftlich abgeleitet*, 2 Vols. Hermann Pohle, Jena. Reprinted by Olms, Hildesheim, 1962; partial English translation Frege (1964).

Frege, G. (1895). Kritische Beleuchtung einiger Punkte in E. Schröder's Vorlesungen über die Algebra der Logik. *Arch. syst. Philos.*, **1**, 433–56. Reprinted in Frege (1967); English translation in Frege (1980*b*, pp. 86–106) and in Frege (1984, pp. 210–28).

Frege, G. (1964). *The basic laws of arithmetic*, (trans. and ed. M. Furth). University of California Press, Berkeley.

Frege, G. (1967). *Kleine Schriften*, (ed. I. Angelelli). Olms, Hildesheim. English translation as Frege (1984).

Frege, G. (1972). *Conceptual notation and related articles*, (ed. T. W. Bynam). Clarendon Press, Oxford.

Frege, G. (1976). *Wissenschaftlicher Briefwechsel*, (ed. G. Gabriel *et al.*). Hamburg. English translation Frege (1980*a*).

Frege, G. (1980*a*). *Philosophical and mathematical correspondence*, (ed. B. McGuiness, trans. H. Kaal). Blackwell, Oxford.

Frege, G. (1980*b*). *Translations from the philosophical writings of Gottlob Frege*, (3rd edn), (ed. P. Geach and M. Black). Blackwell, Oxford.

Frege, G. (1980*c*). *The foundations of arithmetic*, (2nd edn with corrections), (trans. J. L. Austin). Blackwell, Oxford.

Frege, G. (1984). *Collected papers on mathematics, logic, and philosophy*, (ed. B. McGuinness, trans. M. Black *et al.*). Blackwell, Oxford.

Friedman, H. (1971). Higher set theory and mathematical practice. *Ann. Math. Logic*, **2**, 326–57.

Galileo (1638). *Discorsi e dimonstrazioni matematiche, intorno à due nuoue scienze.* Appresso gli Elsevirii, Leida. English translation Galileo (1914) and Galileo (1974).

Galileo (1914). *Dialogues concerning the two new sciences*, (trans. H. Crew and A. de Salvio). New York. Reprinted 1951.

Galileo (1974). *Two new sciences*, (trans. S. Drake). University of Wisconsin Press, Madison.

Gauß, C. F. (1831). Letter to Schumacher, 12 July 1831. In Gauß (1900, p. 216).

Gauß, C. F. (1900). *Werke*, Vol. 8, (ed. Königlichen Gesellschaft der Wissenschaften zu Göttingen). Teubner, Leipzig.

Gödel, K. (1931). Über formal unentscheidbare Sätze der *Principia mathematica* und verwandter Systeme I. *Monatsh. Math. Phys.*, **38**, 173–98. Reprinted in Gödel (1986, pp. 144–94); English translation as Gödel (1962); also translated in van Heijenoort (1967, pp. 596–616); in Davis (1965, pp. 5–38) and in Gödel (1986, pp. 145–95).

Gödel, K. (1938). The consistency of the axiom of choice and the generalized continuum-hypothesis. *Proc. Natl. Acad. Sci. USA*, **24**, 556–7.

Gödel, K. (1944). Russell's mathematical logic. In Schilpp (1944, pp. 125–53). Reprinted with alterations in Benacerraf and Putnam (1983, pp. 447–69); also in Pears (1972, pp. 192–226).

Gödel, K. (1962). *On formally undecidable propositions of Principia Mathematica and related systems.* Oliver & Boyd, Edinburgh.

Gödel, K. (1986). *Collected works*, Vol. I, (ed. S. Feferman *et al.*). Oxford University Press, New York.

Goldblatt, R. (1985). On the role of the Baire category theorem and dependent choice in the foundations of logic. *J. Symb. Logic*, **50**, 412–22.

Goodstein, R. L. (1944). On the restricted ordinal theorem. *J. Symb. Logic*, **9**, 33–41.

Grassmann, H. (1861). *Lehrbuch der Arithmetik für höhere Lehrenstalten.* Easlin, Berlin. Abridged version in Grassmann (1898, pp. 295–349).

Grassmann, H. (1898). *Gesammelte mathematische und physikalische Werke*, Vol. II, (ed. J. Lüroth). Teubner, Leipzig.

Grattan-Guinness, I. (1971). The correspondence between Georg Cantor and Philip Jourdain. *Jahresber. Dtsch. Math.-Ver.*, **73**, 111–30.

Grattan-Guinness, I. (1974). The rediscovery of the Cantor–Dedekind correspondence. *Jahresber. Dtsch. Math.-Ver.*, **76**, 104–39.

Hailperin, T. (1953). Quantification theory and empty individual domains. *J. Symb. Logic*, **18**, 197–200.

Hallett, M. (1984). *Cantorian set theory and limitation of size*, Oxford logic guides, **10**. Clarendon Press, Oxford.

Hardy, G. H. (1904). A theorem concerning the infinite cardinal numbers. *Q. J. Math.*, **35**, 87–94.

Hardy, G. H. (1906). The continuum and the second number class. *Proc. London Math. Soc.* (2), **4**, 10–17.

Hartogs, F. (1915). Über das Problem der Wohlordnung. *Math. Ann.*, **76**, 438–43.

Hausdorff, F. (1908). Grundzüge einer Theorie der geordneten Mengen. *Math. Ann.*, **65**, 435–505.

Hausdorff, F. (1909). Die Graduierung nach dem Endverlauf. *Abh. math.-phys. Kl. königl. sächs. Ges. Wiss.*, **31** [=*Abh. königl. sächs. Ges. Wiss.*, **58**], 295–334.

Hausdorff, F. (1914). *Grundzüge der Mengenlehre.* Veit, Leipzig. Reprinted by Chelsea, New York, 1949; 2nd edn, 1927.

Hayden, S. and Kennison, J. F. (1968). *Zermelo–Fraenkel set theory.* Merrill, Columbus, Ohio.

van Heijenoort, J. (ed.) (1967). *From Frege to Gödel: A source book in mathematical logic, 1879–1931.* Harvard University Press. Cambridge, Mass.

Heine, E. (1872). Die Elemente der Functionlehre. *J. reine angew. Math.*, **74**, 172–88.

Hilbert, D. (1900). Mathematische Probleme. *Nachr. königl. Ges. Wiss. Göttingen, Math.-phys. Kl.*, **7**, 253–97. Reprinted in *Arch. Math. Phys., 3 Reihe*, **1**, (1901), 44–63, 213–37; and in Hilbert (1935, pp. 290–329); English translation in *Bull. Am. Math. Soc.* (*2*), **8**, (1901–2), 437–79; also in *Mathematical developments arising from Hilbert's problems*, (ed. F. E. Browder), Proc. Symp. Pure Math. **28**, Am. Math. Soc., Providence, R.I., 1976, pp. 1–34.

Hilbert, D. (1903). Letter to Frege, 7 November 1903. In Frege (1976, p. 80). English translation in Frege (1980*a*, p. 51).

Hilbert, D. (1925). Über das Unendliche. *Math. Ann.*, **95**, 161–90. English translation in van Heijenoort (1967, pp. 367–92); abridged translation in Benacerraf and Putnam (1983, pp. 183–201); abbreviated version, *Jahresber. Dtsch. Math.-Ver.*, **36**, (1927), 201–15.

Hilbert, D. (1935). *Gesammelte Abhandlungen*, Vol. III. Springer, Berlin.

Hobson, E. W. (1905). On the general theory of transfinite numbers and order types. *Proc. London Math. Soc.* (*2*), **3**, 170–88.

Hodges, W. (1977). *Logic.* Penguin, Harmondsworth.

Isaacson, D. (1987). Arithmetical truth and hidden higher-order concepts. In *Logic Colloquium '85*, pp. 147–69. Elsevier (North-Holland), Amsterdam.

Isbell, J. (1984). Review of 'A compendium of continuous lattices'. *Order*, **1**, 93–5.

Jensen, R. B. (1966). Independence of the axiom of countable dependent choices from the countable axiom of choice. *J. Symb. Logic*, **31**, 294.

Johnstone, P. T. (1987). *Notes on logic and set theory.* Cambridge University Press.

Jourdain, P. E. B. (1905). On a proof that every aggregate can be well-ordered. *Math. Ann.*, **60**, 465–70.

Jourdain, P. E. B. (1906). On the question of the existence of transfinite numbers. *Proc. London Math. Soc.* (*2*), **4**, 266–83.

Kelley, J. L. (1955). *General topology.* Van Nostrand, Princeton, N.J.

Kirby, L. and Paris, J. (1982). Accessible independence results in Peano arithmetic. *Bull. London Math. Soc.*, **14**, 285–93.

Knaster, B. (1927). Une théorème sur les fonctions d'ensembles. *Ann. Soc. Pol. Math.*, **6**, 133–4.

König, J. (1905). Über die Grundlagen der Mengenlehre und das Kontinuumproblem. *Math. Ann.*, **61**, 15–160.

Korselt, A. (1911). Über einen Beweis des Äquivalenzsatzes. *Math. Ann.*, **70**, 294–6.

Kreisel, G. (1967). Informal rigour and completeness proofs. In *Problems in the philosophy of mathematics*, (ed. I. Lakatos), pp. 138–86. North-Holland, Amsterdam.

Kreisel, G. and Levy, A. (1968). Reflection principles and their use for establishing the complexity of axiomatic systems. *Z. math. Logik*, **14**, 97–191.

Kuratowski, K. (1921). Sur la notion d'ordre dans la théorie des ensembles. *Fundam. Math.*, **2**, 161–71.

Kuratowski, K. (1922). Une méthode d'élimination des nombres transfinis des raisonnements mathématiques. *Fundam. Math.*, **5**, 76–108.

Landau, E. (1930). *Grundlagen der Analysis.* Leipzig. English translation as Landau (1951).

Landau, E. (1951). *Foundations of analysis*, (trans. F. Steinhart). Chelsea, New York.

Lennes, N. J. (1922). On the foundations of the theory of sets. *Bull. Am. Math. Soc.*, **28**, 300.

Levi, B. (1902). Intorno alla teoria degli aggregati. *R. Ist. Lombardo Sci. Lett. Rend.* (*2*), **35**, 863–8.

Lévy, A. (1969). The definability of cardinal numbers. In Bulloff *et al.* (1969, pp. 15–38).

Levy, P. (1964). Remarques sur un théorème de Paul Cohen. *Rev. Métaphys. Morale*, **69**, 88–94.

Littlewood, J. E. (1926). *The elements of the theory of real functions, being notes of lectures delivered in the University of Cambridge, 1925.* Heffer, Cambridge.

MacLane, S. (1971). *Categories for the working mathematician*, Graduate Texts in Mathematics **5**. Springer, New York.

MacLane, S. (1986). *Mathematics, form and function.* Springer, New York.

Maier, A. (1949). *Die Vorläufer Galileis im 14. Jahrhundert*, Storia e letteratura, **22**. Rome.

Martin, D. A. (1975). Borel determinacy. *Ann. Math.*, **102**, 363–71.

Mirimanoff, D. (1917). Les antinomies de Russell et de Burali-Forti et le problème fondamental de la théorie des ensembles. *Enseign. Math.*, **19**, 37–52.

Moore, E. H. and Smith, H. L. (1922). A general theory of limits. *Am. J. Math.*, **44**, 102–21.

Moore, G. H. (1982). *Zermelo's axiom of choice: its origins, development, and influence*, Studies in the history of mathematics and physical sciences **8**. Springer, New York.

Morse, A. (1965). *A theory of sets.* Academic Press, New York.

Mostowski, A. (1945). Axiom of choice for finite sets. *Fundam. Math.*, **33**, 137–68.

Mostowski, A. (1948). On the principle of dependent choice. *Fundam. Math.*, **35**, 127–30.

Nachbin, L. (1947). Une propriété charactéristique des algèbres booléiennes. *Port. Math.*, **6**, 115–18.

Nachbin, L. (1949). On a characterization of the lattice of all ideals of a Boolean ring. *Fundam. Math.*, **36**, 137–42.

von Neumann, J. (1923). Zur Einführung der transfiniten Zahlen. *Acta litterarum ac scientiarum Regiae Universitatis Hungaricae Francisco-Josephinae, Sectio scientarum mathematicarum*, **1**, 199–208. Reprinted in von Neumann (1961, pp. 24–33); English translation in van Heijenoort (1967, pp. 346–54).

von Neumann, J. (1925). Eine Axiomatisierung der Mengenlehre. *J. reine angew. Math.*, **154**, 219–40. Reprinted in von Neumann (1961, pp. 34–56); English translation in van Heijenoort (1967, pp. 393–413).

von Neumann, J. (1961). *Collected works*, Vol. 1. Pergamon, Oxford.

Noether, E. and Cavaillès, J. (1937). *Briefwechsel Cantor–Dedekind*, Actualités scientifiques et industrielles, **518**. Hermann, Paris.

Peano, G. (1889). *Arithmetices principia nova methodo exposita*. Turin. Reprinted in Peano (1957–9, Vol. 2, pp. 20–55); partial English translation in van Heijenoort (1967, pp. 83–97) and in Peano (1973, pp. 101–34).

Peano, G. (1890). Démonstration de l'intégrabilité des équations différentielles ordinaires. *Math. Ann.*, **37**, 182–228. Reprinted in Peano (1957–9, Vol. 1, pp. 119–70).

Peano, G. (1906). Super theorema de Cantor–Bernstein. *Rend. Circ. Mat. Palermo*, **21**, 360–6. Reprinted in *Riv. Mat.*, **8**, (1906), 136–43 and in Peano (1957–9, Vol. 1, pp. 337–44).

Peano, G. (1957–9). *Opera scelte*, 3 Vols, (ed. U. Cassina). Cremonese, Rome.

Peano, G. (1973). *Selected works of Guiuseppe Peano*, (trans. and ed. H. C. Kennedy). Allen & Unwin, London.

Pears, D. F. (ed.) (1972). *Bertrand Russell: a collection of critical essays*. Anchor, Garden City, N.Y.

Peirce, C. S. (1881). On the logic of number. *Am. J. Math.*, **4**, 85–95. Reprinted in Peirce (1933, pp. 158–70).

Peirce, C. S. (1933). *Collected papers*, Vol. III, (ed. C. Hartshorne and P. Weiss). Harvard University Press, Cambridge, Mass.

Poincaré, H. (1906). Les mathématiques et la logique III. *Rev. Métaphys. Morale*, **14**, 294–317.

Purkert, W. and Ilgauds, H. J. (1987). *Georg Cantor: 1845–1918*, Vita Mathematica, **1**. Birkhäuser, Basle.

Quine, W. V. O. (1941). *Mathematical logic*. Harvard University Press, Cambridge, MA. Revised edn, 1951; reprinted by Harper, New York, 1962.

Quine, W. V. O. (1953). *From a logical point of view*. Harvard University Press, Cambridge, Mass. 2nd edn, 1963.

Quine, W. V. O. (1954). Quantification and the empty domain. *J. Symb. Logic*, **19**, 177–9.

Quine, W. V. O. (1963). *Set theory and its logic*, Harvard University Press, Cambridge, Mass.

Ramsey, F. P. (1926). The foundations of mathematics. *Proc. London Math. Soc.* (*2*), **25**, 338–84. Reprinted in Ramsey (1931, pp. 1–61).

Ramsey, F. P. (1931). *The foundations of mathematics and other logical essays*. Kegan Paul, Trench, Trubner, London.

Rang, B. and Thomas, W. (1981). Zermelo's discovery of Russell's paradox. *Hist. Math.*, **8**, 15–22.

Robinson, A. (1939). On the independence of the axioms of definiteness (Axiome der Bestimmtheit). *J. Symb. Logic*, **4**, 69–72.

Rosser, J. B. (1953). *Logic for mathematicians*. McGraw-Hill, New York.

Rubin, H. (1960). Two propositions equivalent to the axiom of choice only under both the axioms of extensionality and regularity. *Not. Am. Math. Soc.*, **7**, 381.

Rudin, M. E. (1971). The box topology. In *Proc. Univ. Houston Point Set Topology Conf.*, pp. 191–9. Houston, Tex.

Russell, B. (1903). *The principles of mathematics*. Allen & Unwin, London. 2nd edn, 1937.

Russell, B. (1906). On some difficulties in the theory of transfinite numbers and order types. *Proc. London Math. Soc.* (*2*), **4**, 29–53. Reprinted in Russell (1973, pp. 135–64).

Russell, B. (1908). Mathematical logic as based on the theory of types. *Am. J. Math.*, **30**, 222–62. Reprinted in Russell (1956, pp. 59–102); also in van Heijenoort (1967, pp. 150–82).

Russell, B. (1919). *Introduction to mathematical philosophy.* Allen & Unwin, London.

Russell, B. (1936). The limits of empiricism. *Proc. Aristotelian Soc., NS*, **36**, 131–50.

Russell, B. (1944). My mental development. In Schilpp (1944, pp. 3–20).

Russell, B. (1956). *Logic and knowledge, Essays 1901–1950*, (ed. R. C. Marsh). Allen & Unwin, London.

Russell, B. (1973). *Essays in analysis*, (ed. D. Lackey). Braziller, New York.

Schilpp, P. A. (ed.) (1944). *The philosophy of Bertrand Russell*, Library of living philosophers, **5**. Northwestern University, Evanston, Ill.

Schönflies, A. (1905). Über wohlgeordnete Mengen. *Math. Ann.*, **60**, 181–6.

Schönflies, A. (1913). *Entwickelung der Mengenlehre und ihrer Anwendungen.* Teubner, Leipzig.

Schröder, E. (1898). Über zwei Definitionen der Endlichkeit und G. Cantor'sche Sätze. *Nova Acta Academiae Caesareae Leopoldino-Carolinae Germanicae Naturae curiosorum* [= *Abhandlungen der Kaiserlichen Leopoldinischen-Carolinischen Deutschen Akademie der Naturforscher*], **71**, 303–62.

Scott, D. (1955). Definitions by abstraction in axiomatic set theory. *Bull. Am. Math. Soc.*, **61**, 442.

Scott, D. (1962). More on the axiom of extensionality. In *Essays on the foundations of mathematics, dedicated to A. A. Fraenkel on his seventieth birthday*, (ed. Y. Bar-Hillel, I. J. Poznanski, M. O. Rabin, and A. Robinson), pp. 115–31. North-Holland, Amsterdam.

Scott, D. (1974). Axiomatizing set theory. In *Axiomatic set theory*, Proc. Symp. Pure Math. **13**, Part II, pp. 207–14. Am. Math. Soc., Providence, R.I.

Sierpinski, W. (1918). L'axiome de M. Zermelo et son rôle dans la Théorie des Ensembles et l'Analyse. *Bulletin International de l'Académie des Sciences de Cracovie, Classe des Sciences Mathématiques et Naturelles, Série A: Science Mathématiques*, 97–152. Reprinted in Sierpinski (1975, pp. 208–55).

Sierpinski, W. (1924). Sur l'hypothèse du continu ($2^{\aleph_0} = \aleph_1$). *Fundam. Math.*, **5**, 177–87. Reprinted in Sierpinski (1975, pp. 527–36).

Sierpinski, W. (1975). *Oeuvres choisies*, Vol. II, (ed. S. Hartman *et al.*). Panstwowe Wydawnictwo Naukowe, Warsaw.

Sinaceur, M. A. (1971). Appartenance et inclusion: un inédit de Richard Dedekind. *Rev. Hist. Sci.*, **24**, 247–54.

Skolem, T. (1922). Einige Bemerkungen zur axiomatischen Begründung der Mengenlehre. *Wiss. Vorträge gehalten auf dem 5 Kongress der Skandinav. Mathematiken in Helsingfors*, pp. 217–32. English translation in van Heijenoort (1967, pp. 290–301).

Steinitz, E. (1910). Algebraische Theorie der Körpern. *J. reine angew. Math.*, **137**, 163–309.

Stone, M. H. (1936). The theory of representations for Boolean algebras. *Trans. Am. Math. Soc.*, **40**, 37–111.

Stone, M. H. (1937*a*). Topological representation of distributive lattices and Bourwerian logics. *Cas. Pest. Mat. Fys.*, **67**, 1–23.

Stone, M. H. (1937*b*). Applications of the theory of Boolean rings to gener topology. *Trans. Am. Math. Soc.*, **41**, 375–481.

Szpilrajn, E. (1930). Sur l'extension de l'ordre partiel. *Fundam. Math.*, **16**, 386–9

Tarski, A. (1924). Sur quelques théorèmes qui équivalent à l'axiome du choi *Fundam. Math.*, **5**, 147–54.

Tarski, A. (1955). The notion of rank in axiomatic set theory and some of it applications. *Bull. Am. Math. Soc.*, **61**, 443.

Teichmüller, O. (1939). Braucht der Algebraiker das Auswahlaxiom? *Dtsche. Math* **4**, 567–77.

Tietze, H. (1923). Beitrage zur allgemeinen topologie I. *Math. Ann.*, **88**, 280–31

Tukey, J. W. (1940). *Convergence and uniformity in topology*, Annals of Mathe matics Studies **2**. Princeton University Press.

Tychonoff, A. N. (1929). Über die topologische Erweiterung von Räumen. *Mat Ann.*, **102**, 544–61.

Tychonoff, A. N. (1935). Über einen Funktionenräum. *Math. Ann.*, **111**, 762–6.

van der Waerden, B. L. (1930). *Moderne Algebra*, 2 Vols. Springer, Berlin.

Weyl, C. H. H. (1910). Über die Definitionen der mathematischen Grundbegriff *Math. Naturwiss. Bl.*, **7**, 93–5, 109–13. Reprinted in Weyl (1968, pp. 298–304)

Weyl, C. H. H. (1949). *Philosophy of mathematics and natural science*. Princeto University Press.

Weyl, C. H. H. (1968). *Gesammelte Abhandlungen*, Vol. I, (ed. K. Chandrasekhran Springer, Berlin.

Whitehead, A. N. (1902). On cardinal numbers. *Am. J. Math.*, **24**, 367–94.

Whitehead, A. N. and Russell, B. (1910–13). *Principia mathematica*, 3 Vols. Cam bridge University Press. 2nd edn, 1925–7.

Wiener, N. (1914). A simplification of the logic of relations. *Proc. Cambridg Philos. Soc.*, **17**, 387–90. Reprinted in van Heijenoort (1967, pp. 224–27).

Wittgenstein, L. (1922). *Tractatus logico-philosophicus*. Kegan Paul, Trench Trubner, London.

Zermelo, E. (1904). Beweis, dass jede Menge wohlgeordnet werden kann. *Mat Ann.*, **59**, 514–16. English translation in van Heijenoort (1967, pp. 139–41)

Zermelo, E. (1908*a*). Neuer Beweis für die Möglichkeit einer Wohlordnung. *Mat Ann.*, **65**, 107–28. English translation in van Heijenoort (1967, pp. 183–98).

Zermelo, E. (1908*b*). Untersuchungen über die Grundlagen der Mengenlehr I. *Math. Ann.*, **65**, 261–81. English translation in van Heijenoort (1967 pp. 199–215).

Zermelo, E. (1930). Über Grenzzahlen und Mengenbereiche. *Fundam. Math.*, **16** 29–47.

Zorn, M. (1935). A remark on method in transfinite algebra. *Bull. Am. Math. Soc* **41**, 667–70.

Index of notation

Index of terminology